Ernest Zebrowski, Jr. Community College of Beaver County
formerly of Westinghouse Research Laboratories and Jones and Laughlin Steel Corporation Research Laboratories

FUNDAMENTALS OF

Physical Measurement

Duxbury Press
North Scituate, Massachusetts

Duxbury Press
A Division of Wadsworth Publishing Company, Inc.

©1979 by Wadsworth Publishing Company, Inc., Belmont, California 94002

All rights reserved. No part of this book may be reproduced, stored in a retrieval system, or transcribed, in any form or by any means, electronic, mechanical, photocopying, recording, or otherwise, without the prior written permission of the publisher, Duxbury Press, a division of Wadsworth Publishing Company, Inc., Belmont, California.

Fundamentals of Physical Measurement
was edited and prepared for composition by Penelope Rohrbach.
Interior design was provided by Amato Prudente.
The cover was designed by Joe Landry.

Library of Congress Cataloging in Publication Data

Zebrowski, Ernest, Jr.
  Fundamentals of physical measurement.

  Includes index.
  1. Physical measurement I. Title.
QC39.Z4   530'.8   78-7315
ISBN 0-87872-173-8

L.C. Cat. Card No.: 78-7315
ISBN 0-87872-173-8

Printed in the United States of America
1 2 3 4 5 6 7 8 9–82 81 80 79 78

CONTENTS

	Preface	v
1.	Units and Standards	1
2.	Measurement Error	14
3.	Dealing with Uncertainties	25
4.	Accuracy versus Precision	50
5.	Data Tables	65
6.	Construction of Graphs	77
7.	The Propagation of Uncertainty	100
8.	Empirical Equations	124
9.	Polar Graphs	156
10.	Logarithmic Graphs	171
11.	Indicators and Recorders	190
12.	Sensors	213

13. Dimensional Analysis and the International System   **235**

Appendix A   Tables of Conversion Factors   253

Appendix B   The International Systems of Units (SI)   283

Index   291

PREFACE

Measurement plays a fundamental role in our modern world. In commerce, goods are priced by volume, mass, or sometimes length or area; services such as transportation are billed according to quantity of material as well as according to the distance it is transported. In commercial transactions, errors in measurement have a direct bearing on profits and costs.

In the engineering technologies, every project begins and ends with measurements. The design of a highway or skyscraper starts with a survey; the design of a power transformer starts with measurements of the electrical and magnetic properties of the wire, the insulation, and the magnet core. The final product must then be tested to see if it actually measures up to its theoretical performance. A recent near-disaster at a power plant was traced to a technician's certification that a certain pipe's concentricity was within specifications. It was not, and it subsequently failed. Replacement required direct costs, for materials and labor, of some $50,000. Indirect costs, due to power that was not produced, amounted to several *million* dollars.

Even more serious than potential financial loss, however, is the threat to life and limb that can result from inaccurate measurement. A dam failure of not too long ago occurred only hours after the structure had been inspected and declared safe. The two inspectors were among those who perished.

Perhaps less dramatic, but nonetheless important for those concerned, is measurement's role in the natural sciences. Science has progressed to the point where new discoveries tend to be very subtle refinements of previously established principles. A scientist is not likely to stumble over a new phenomenon as if it were a cat sleeping on the doorstep. Today's scientific advances are based on the comparison of precise measurements with the predictions of existing theory. The discovery of the solar wind and the radioactive fallout of the 1950s are well-publicized examples.

Over the past 10 years, it has become increasingly apparent to me that science, engineering, and technical students need an early and coherent introduction to the subject of measurement. I have heard too many beginning students blame a disappointing experimental result on "human error" or "lousy equipment"; I have seen too many fail to correct for obvious systematic errors, neglect to control extraneous variables, then draw graphs that look like a chart of the daily fluctuation of the Dow-Jones Average. I have watched too many students neglect to exploit an instrument's full precision, then pick up a calculator and quote an 8-digit result. And I have puzzled too often over data tables without units, graphs without legends, and calculated results without error limits. True, there are a number of existing texts on the statistical treatment of experimental data. There are also a few devoted to the specific problems of experimental design. None, however, give a complete overview of the general problem of measurement—beginning with standards and the International System (SI), progressing through the identification of errors, the propagation of uncertainty in indirect measurements, the details of table and graph construction, the derivation of empirical equations, hardware considerations in the selection of sensors and indicators, and dimensional analysis with SI units. There are also none that, in my estimation, treat any of these topics at a truly introductory level.

It is the purpose of this book, then, to provide an overview devoid of the statistical complications that are well documented elsewhere. For students in technical or business programs, this should be sufficient. For students of science and engineering, it should provide a perspective against which the more specialized analytical details can be viewed later.

To keep the mathematics from obscuring the principles, none of the procedures herein require more than simple algebra. The few necessary statistics are described in detail and, through the use of Peter's approximation, circumvent the tedium of root-mean-square calculations. Although a form of statistical regression is needed to write empirical equations, the least-squares technique is avoided through a simple extension of this same procedure. This has the added advantage of giving the identical regression curve for x on y as for y on x, a necessary but much-neglected prerequisite to any algebraic transposition of the variables in the empirical equation.

All important definitions are set apart from the main text and identified as such, and all important procedures are illustrated through solved examples. Sets of review questions and exercises are included at the end of each chapter. The instructor may wish to consider the questions as a set of behavioral objectives.

Answers to approximately half the problems are included with the problems themselves. This allows students to confirm that they are on the right track, while leaving the instructor some problems to assign for credit if he or she wishes.

The appendixes include some 600 unit conversion factors, distinguished according to those that are exact by international definition and those that themselves are the result of measurement. Also included is a detailed guide to the International System.

Finally, a word of thanks is in order to the many students who patiently tested the pedagogical effectiveness of this material. I sincerely hope I have done them some small service.

Ernest Zebrowski, Jr.

ACKNOWLEDGMENTS

The following companies and governmental agencies have generously contributed technical information and illustration materials for use in this book.

American Motors Corporation
Bacharach Instrument Company
The Bendix Corporation
Brookfield Engineering Laboratories, Inc.
Central Tool Company, Inc.
Chrysler Corporation
Daedalon Corporation
Detroit Testing Machine Company
Federal Aviation Administration
Fischer & Porter Company
Ford Motor Company
General Motors Corporation
Heath Company
Heidelberg Eastern, Inc.
Hewlett-Packard
Honeywell
Keuffel & Esser Company
Lawrence Berkeley Laboratory
Markson Science Inc.
Mettler Instrument Corporation
National Bureau of Standards
Ohaus Scale Corporation
Oriel Corporation
RCA
Sargent-Welch Scientific Company
J. T. Slocomb Company
The L. S. Starrett Company
Taylor Instrument Corporation
Tektronix, Inc.
Union Camp Corporation
United States Steel Corporation
Vacuum Accessories Corporation of America
Voland Corporation
Westinghouse

CHAPTER 1

Units and Standards

All of us are used to solving everyday problems through arithmetic and possibly through algebra. We balance our checkbooks, estimate our driving time to a distant city, calculate how many shingles we need to replace a roof. We are probably so used to doing these problems that we automatically think of mathematics as a tool that exists expressly for such purposes. But that is not really the case.

The Role of Mathematics

Mathematics is a scheme for dealing with numbers and with functions, or sets of numbers. This scheme tells us how to operate on numbers to get new numbers, and how to operate on functions to get new functions. But mathematics itself does not tell us what these numbers or functions mean, as far as anything physical is concerned.

Numbers and mathematical functions acquire physical meaning only when engineers and scientists become involved. Their job is to find ways of expressing properties of the real world in numerical form. Only then does mathematics become a practical tool. And if the expressions are translated properly, mathematics can be a tool just as valuable as any wrench or soldering gun. For example, through calculation, we can determine whether a crane platform will be stable before the platform is actually built; we can predict the performance of a lunar vehicle before it leaves the earth; we can estimate the fuel needs for a building, based on the insulating properties of its walls and windows; or we can analyze the wearing of piston rings in an engine, based on the buildup of radioactive tracers in the oil.

Counting

Obviously, such calculations are useful only if the numbers have a valid physical meaning. We can attach a *physical* significance to a number in two ways. First, we can use the number to represent the result of a *count* of a set of objects. For instance, if a car engine needs eight spark plugs, we can easily see if we have enough replacements simply by counting to eight. "Eight" now represents not just a number, but a number of physical objects. Although we all learned this simple concept in elementary school, it is mentioned here only to point out its distinction from the second way we relate numbers to the real world.

Measuring

That second way is to use the number to represent the result of a *measurement*. Suppose, for example, that we need to know the height of a certain radio tower. Using a theodolite or a sextant, we might determine the height to be 297 feet above ground level. But we are not using 297 as a counting number; we do not have exactly 297 things called "feet." Instead, we have indirectly compared the height of the tower with the height of something we have designated as 1 "foot." And we have concluded that it would take approximately 297 of those feet to make up the tower's height.

Comparisons and Units

Measurement is a process of comparison. Had we compared the radio tower's height to something we call a "yard," the result of the measurement would have been 99 instead of 297. Had we compared it to something we call a "metre," the result would have been 90.5. Had we used inches, the measurement's numerical result would have been 3560.

Obviously, it is just as important to specify the thing we are comparing with as it is to quote the number itself. This is why we say that the numbers by themselves are meaningless. In the technologies, it is absolutely essential that we have a *unit* associated with each number we use. You will notice that units are specified for all numbers throughout this book.

We will use many different kinds of units: for example, units of distance (inches, centimetres, feet, yards, metres, kilometres, miles, and so on) and units of time (milliseconds, seconds, minutes, hours, days, years, etc.). In addition, we use combinations of units, or compound units, for some quantities: speed, for instance, may be expressed in feet per second, kilometres per hour, miles per hour, or any of a number of other combinations. And *every* time we make a measurement, we are comparing the known size of one of these units with the size of the quantity we are measuring. Even if a comparison must be made indirectly, it is still a comparison. It is important to keep this point in mind whenever undertaking measurements.

Standards

How do we keep track of the exact size of each unit? The unit is the thing that has to have some relation to the physical world, so just keeping track of its size on paper will not work. Instead, we need to use physical records, called *standards*, to permanently record the size of units.

DEFINITION | *A standard is a permanent or readily reproducible physical record of the size of a unit of measurement.*

If, for example, we were trying to accurately measure the electrical resistance of a coil of wire, we would need a standard resistor whose resistance was known very accurately. This standard resistor would be our physical record of the size of the unit of resistance: the *ohm,* abbreviated Ω (capital Greek omega). The standard resistor need not be exactly 1 ohm; it might be, say, 5.79 ohms. The point is that the unit's size is permanently recorded in the resistance characteristics of a physical object. The measurement then amounts to comparing the unknown resistance with our standard, perhaps by using a circuit known as a Wheatstone bridge. If the unknown turns out to be 9.81 times as big as the standard, then the result becomes

$$(9.81)(5.79 \, \Omega) = 56.8 \, \Omega$$

"Wait a minute!" you say. "I never use a standard resistor to measure resistance. I just take my pocket ohmmeter, clip on the leads, and read the scale!" True, that might be what you do. But an ohmmeter actually has its own standard resistors built in (in fact, a separate resistor for each range). The ohmmeter still does what any measuring instrument does: it compares an unknown with a standard. The standard always has to come in somewhere.

What about a surveyor's transit? Shown in figure 1-1, this device is used to make precise angular measurements. It has a very finely ruled circular scale, which, through the use of a vernier (see p. 4), is divided into minutes of an arc. This scale acts as the standard. The rest of the instrument—the rotating telescope with its cross hairs—compares the positions of landmarks in the field with the divisions on the standard scale.

What about a thermometer? A thermometer is an instrument for measuring temperature, yet no standard temperatures are built into it. And certainly we do not have to keep a standard temperature on hand when we use one. But even so, we cannot measure temperatures without dealing with a standard. When

Figure 1-1. *A surveyor's transit. Standards for angular measurement are built into the instrument. (Courtesy of Keuffel & Esser Company)*

a thermometer is manufactured, it is *calibrated* against a standard, sometimes by placing it in a bath known to be 0°C, then adjusting the scale so it actually reads 0°C. After that, every time we use the thermometer, we are actually comparing our unknown temperature with the temperature of the manufacturer's standard. There is just no way to avoid using standards.

Primary and Secondary Standards

But how are all these standards created? Do we need standards for our standards, to make sure they are all the same? Yes. We have both *primary* and *secondary* standards.

DEFINITION | A <u>primary standard</u> is used as a fundamental definition of the size of a physical unit. Primary standards are specified by the provisions of an international treaty.

DEFINITION | A <u>secondary standard</u> is a copy of a primary standard that can be used routinely for making measurements.

Few of us will ever see a primary standard, let alone use one. Rather, we will generally deal with a secondary standard that has been copied from another secondary standard that itself may be many steps removed from the primary standard. Occasionally we hear about *laboratory standards,* which are secondary standards certified to be especially accurate reproductions of primary standards. We must remember that our measurements will never be better than our standards; in fact, usually they will be a good bit worse. But to have extreme accuracy, it is necessary to work with accurate standards.

Before 1960, most developed countries maintained their own sets of primary standards. In the United States, the standards were kept at the National Bureau of Standards near Washington, D.C.; in Britain, they were kept at the National Physical Laboratory at Teddington. Naturally, these standards varied slightly from one nation to another. This variation caused considerable problems for scientists and technicians.

The International System

In 1960, 36 nations signed a treaty and created an international system of units based on only *one* set of primary standards, which were housed in Sèvres, France. The entire system is officially called "Le Système International d'Unités" (SI), and is called the *International System* in English-speaking countries. The SI is essentially the same as what we have come to know as the *metric system.*

Every six years, an International Committee on Weights and Measures (CGPM) meets to recommend improvements to this system. The most important improvement has been to make the primary standards more available to scientists and technicians around the world. This has been done by making certain precisely reproducible physical phenomena—rather than physical artifacts—the basis of the standards. At first, for instance, the metre was defined as the distance between two fine lines engraved on gold plugs near the ends of a certain bar made of a platinum-iridium alloy.* The bar had to be at a temperature of 0.00°C and supported mechanically in a prescribed way. The bar was not very accessible, and making accurate copies was difficult. Today the metre is defined in terms of the wavelength of the light from a krypton-86 light source. Although this light source is very expensive, anyone who really needs a primary standard of length can now have his or her own.

---

*Metre is sometimes also spelled *meter. Metre* is preferred because it avoids confusion with the meter that is a measuring instrument.

Base Units

Table 1-1 lists the seven SI base units and their primary standards. Although some of these quantities and their standards may have little meaning to you right now, this should be no cause for alarm. What you need to appreciate

Table 1-1. *The SI Base Units and Their Primary Standards*

Quantity	Unit	Primary Standard
length	metre (m)	1 650 763.73 wavelengths of the light from a certain transition in a krypton-86 atom
time	second (s)	the duration of 9 192 631 770 periods of the radiation from a certain transition in a cesium-133 atom
mass	kilogram (kg)	the mass of a particular platinum-iridium cylinder at the International Bureau of Weights and Measures at Sèvres, France
electric current	ampere (A)	the current that, if maintained in two very long, parallel conductors of negligible cross section and placed 1 metre apart in vacuum, will produce between those conductors a force equal to 2×10^{-7} newtons per metre of length
temperature	kelvin (K)	the fraction 1/273.16 of the thermodynamic temperature of the triple point of water
luminous intensity	candela (cd)	the luminous intensity emitted from an area of $1/60$ square centimetre of molten platinum at 2 045 K
amount of substance	mole (mol)	the amount of substance of a system that contains as many elementary entities as there are atoms in 0.012 kilogram of carbon-12

Note: This book follows the CGPM recommendation that many-digit numbers be divided into groups of three digits in both directions from the decimal indicator. The groups are separated by a space rather than a comma.

now is that the International System does exist, together with a well-defined set of primary standards, and that a great deal of work and international cooperation have gone into its creation. One thing that all this work has produced is the realization that there are in fact only seven fundamentally different physical quantities that can be measured. The seven base units are the units for these seven quantities.

DEFINITION | *The SI base units are the official international units for the seven different kinds of physical quantities that can be measured.*

We can measure length, which includes width, height, thickness, distance, and so on. We can measure time, which is fundamentally different. We can measure mass, electric current, temperature, luminous intensity, and amount of substance. We can measure quantities that are combinations of some of these fundamental seven, but no one has ever encountered a measurement that cannot be referred to these seven fundamental quantities and their base units.

Of these base units, only one, the *kilogram*, is presently defined in terms of an artifact—that is, a single physical sample. The kilogram's primary standard is a certain platinum-iridium cylinder kept at the International Bureau of Weights and Measures in France. Copies of this cylinder have been distributed to the national laboratories of the nations who signed the Treaty of the Metre. The United States prototypes are shown in figure 1-2. Every mass and weight measurement made in the United States is an indirect comparison with one of the standards shown.

The primary standards for other base units require elaborate laboratory setups. Figure 1-3, for instance, shows a current balance at the National Bureau

Figure 1-2. *The U.S. standards for the kilogram, the only unit now defined by an artifact. (Courtesy of National Bureau of Standards)*

Figure 1-3. *A current balance provides a readily reproducible standard for electric current. (Courtesy of National Bureau of Standards)*

of Standards. This instrument is used to produce an electric current consistent with the definition of the ampere in table 1-1.

The International Practical Temperature Scale

Furthermore, advances in measurement technique simply have not yet caught up with some of the definitions. This is particularly the case with the kelvin (K), the SI unit for temperature. The primary standard is based on the triple point of water (the temperature and pressure at which water will simultaneously boil and freeze). Finding the triple point in the laboratory is not particularly difficult; the problem lies in finding 1/273.16 of this temperature. Since no one has discovered an accurate way to do so, we have to rely instead on a series of temporary practical standards (table 1-2). These standards are readily reproducible, but they are not as accurate as having a single primary standard for temperature.

You will also notice, in tables 1-1 and 1-2, that numbers with many digits are written in a special way. The old U.S. procedure was to use commas to separate very large numbers into groups of three digits. In many metricated countries, however, the comma is used as the decimal indicator; a French scien-

Table 1-2. *Fixed Points of the International Practical Temperature Scale* (IPTS)

Substance and Equilibrium State	Assigned Temperature (K)
1. Gold (solid/liquid)	1 337.58
2. Silver (solid/liquid)	1 235.08
3. Zinc (solid/liquid)	692.73
4. Tin (solid/liquid)	505.118 1
5. Water (liquid/vapor)	373.15
6. Water (triple point)	273.16
7. Oxygen (liquid/vapor)	90.188
8. Oxygen (triple point)	54.361
9. Neon (liquid/vapor)	27.102
10. Hydrogen (liquid/vapor)	20.28
11. Hydrogen (liquid/vapor, at a pressure of 0.329 standard atmosphere)	17.042
12. Hydrogen (triple point)	13.81

tist will write 273,16 to mean what a U.S. scientist means by 273.16. This leads to trouble with a number written as 9,192,631,770, for instance. To avoid confusion, the CGPM recommends against the use of the comma in numbers as anything but a decimal indicator, at least in printed works. Instead, the groups of three digits are separated by a space, so the number just mentioned becomes 9 192 631 770. Furthermore, the space is inserted between groups of three digits on both sides of the decimal indicator. The number of kilometres in a statute mile, for instance, is written as 1.609 344 . Since this book was written in the United States, it still uses the decimal point as the decimal indicator.

Defined Units

Of course, we use many units other than the base units listed in table 1-1. We might ask, for instance, what the standard is for the *inch*. The answer is that the inch has no primary standard, since by international agreement the primary unit of length is the metre. But regardless of international agreement, some of us still like to measure lengths in inches. So for us diehards, the U.S. inch has been defined to be a standard fraction of the international metre—specifically, 0.025 40 of a metre. There are many other such *defined units;* see the tables in appendix A (p. 249) for their numerical relationships. You are certain to find them useful in the future.

DEFINITION | *Defined units are not part of the International System. Rather, they have been related to the corresponding SI units through numerical definitions.*

Compound or Derived Units

Units that can be expressed as combinations of the SI base units are called *compound units* or *derived units*.

DEFINITION | *Compound, or derived, units are units for all quantities other than length, time, mass, electric current, temperature, luminous intensity, and amount of substance. In the International System (SI), all such units can be expressed as a combination of some of the base units.*

The compound units of the International System are listed in appendix tables B-3, B-7, and B-9. Although compound units themselves have no primary standards, the units that compose them do. The SI unit for speed, for instance, is the *metre per second*. There is one primary standard for the metre, and another one for the second. Used together, they give us the standard for the metre per second. Similarly, using miles per hour presents no real problem. Appendix A gives the defined relation between the mile and the metre, and we know that there are 3 600 seconds in an hour. So no matter what measurement we make, we are always indirectly working our way back to comparisons with the primary SI standards.

Calibration

Earlier, mention was made about calibrating measuring instruments. Any instrument that does not carry its own secondary standard internally will have to be calibrated periodically. With the increasing sophistication of modern instrumentation, proper calibration has become very important and sometimes challenging.

DEFINITION | *Calibration is the process of bringing a measuring instrument into agreement with a physical standard or set of standards.*

A voltmeter, for instance, carries no voltage standard of its own. As time passes, heat and atmospheric conditions cause its electronic components to age and change their electrical properties slightly. As a result, the voltage readings

Figure 1-4. *A technician calibrates a piece of electronic equipment. With the sophistication of modern instrumentation, calibration can be exacting. (Courtesy of Tektronix, Inc.)*

become less and less accurate, and it is impossible to predict whether they are getting too high or too low. In fact, there is no way to tell if the calibration is necessary until it is actually performed. The technician who complains that an instrument sent in for calibration actually needs little or no adjustment has missed the point. If no adjustment is necessary, the user can place confidence in the last set of measurements. And it means that the instrument is not subject to wild variations in accuracy. Of course, if a very important set of measurements is to be made, prudence would suggest that the calibration be done *first*.

Unfortunately, the process of calibration is different for every instrument. Figure 1-4 shows a technician calibrating some electronic equipment. The manufacturer's operating manual usually gives a step-by-step procedure for the particular instrument, and also lists the exact physical standards needed for the calibration. Thus it is a good idea to catalogue and file all manufacturer's operating manuals, and to replace immediately any that have been lost or damaged.

Nature has the habit of working changes on all physical objects, measuring instruments (and people) included. Electronic devices are particularly susceptible to these changes, and we should always take the time to go through the calibration procedures when accuracy is important. Mechanical devices are susceptible to friction, wear, and sometimes fatigue. Mechanical instruments that depend on weights are subject to weight gains through oxidation. Relatively minor temperature fluctuations may affect distance-measuring equipment. Optical instruments are affected by temperature, mechanical vibration, and occasionally (as with replica diffraction gratings) humidity. We have to be constantly aware that these changes are taking place to some extent, regardless of the care taken to prevent them. And the only way to place any confidence in the result of a measurement is to make sure that the instrument has been recently and properly calibrated against a reliable standard.

Summary

A number can have physical meaning either by representing a *count* of a set of objects or by representing the result of a measurement. *Measurement* is a comparison of a physical quantity with a *unit*. The size of each unit is recorded in a permanent or readily reproducible *standard*. The International System of Units (SI) is the official set of internationally acceptable units and their primary standards. All defined and derived units can be referred to that fundamental system. Every measuring instrument must be *calibrated* against a standard that can be referred to the primary standards of the International System.

REVIEW QUESTIONS

1. Why is it important to be able to describe things in numerical terms?

2. What are the two ways to give physical meaning to a number?

3. Measurement is said to be a process of comparison. What is being compared when you do each of the following?
   a. weigh yourself on a scale
   b. read a stopwatch at a track and field event
   c. check a spark plug gap with a feeler gauge
   d. read an automobile's speedometer
   e. measure your waistline with a tape measure

4. What is a unit?

5. What is a standard?

6. What is the difference between a primary standard and a secondary standard?

7. What did the treaty of 1960 accomplish?

8. What is the standard abbreviation for the official international version of the metric system?

9. What is a base unit? Give an example.

10. What is a defined unit? Give an example.

11. What is a compound, or derived, unit? Give an example.

12. Name the fundamental physical quantities.

13. Name the base units in the International System.
14. What is calibration?
15. Why is it necessary to calibrate instruments periodically?

EXERCISES

1. Make a list of all the units you can think of for measuring length. Compare your list with the list in appendix tables A-1 and A-2.
2. Make a list of all the units you can think of for measuring area. Compare your list with the list in appendix A, table 2.
3. Make a list of all the units you can think of for measuring volume. Compare your list with appendix A, table 3.
4. In the following list, identify the units.

   length   pressure   mole
   bushel   pound   mass
   weight   kelvin   luminous intensity
   candela   area   miles per hour

5. In the following list, identify the physical quantities.

   inch   electric current   metre
   gallon   kelvin   time
   length   temperature   acre
   speed   furlong   volume

6. In the following list, identify the defined units.

   ampere   gallon   kelvin
   foot   pound   candela
   mole   kilogram   mile

7. In the following list, identify the derived units.

   kilogram   metre per second
   ampere   cubic metre
   volt   watt
   kilogram per cubic metre   second
   mole per second   candela per square metre

CHAPTER 2

Measurement Error

Right and Wrong

The solution to a purely mathematical problem is either right or wrong. Either we have properly followed the rules of mathematics in working out our answer, or else we have slipped up somewhere. In pure mathematics, we seldom talk about answers as being "nearly right" or only "slightly wrong."

When we make a measurement, however, the situation is quite a bit different. In one sense the result of every measurement we make is wrong, since our instruments and secondary standards are both subject to physical limitations that prevent absolute accuracy. Later we will spend more time discussing the origin of these limitations, but for now you need only know that they exist. And because they do, no measurement will ever yield the exact "true value" of the quantity being measured.

This situation is not really as dismal as it might sound. No one really cares if a public right-of-way on his or her property is off by a few inches one way or the other. No one cares if the flash point of a new organic solvent has been reported incorrectly by a few tenths of a degree. No one cares if the voltage of a 12-volt battery is really 12.21 volts. But if the right-of-way is off by 50 feet, or the flash point is off by 60 degrees, or the battery voltage is off by 7 volts, then certainly some people will care very much.

Measurement Error

So we cannot just throw up our hands and say, "What the heck, measurements are always wrong anyway!" What we have to do is examine *how* wrong we can allow them to be. And instead of calling them wrong, we will talk about the *measurement error*, and how big this error is likely to be.

DEFINITION | <u>Measurement error</u> is the difference between a physical quantity's "true value" and its measured value.

Notice that we are now using the word *error* in a very special, technical sense. In the sciences and technologies, an error is not the same thing as a mistake. Mistakes can be avoided, or at least corrected. But errors in measurement can never be eliminated completely. The best we can do is to try to keep the errors small enough that the result can still be used for its intended purposes. In other words, we must define the limits of acceptable error.

Limits of Acceptable Error

▶ **Example 2-1: The speedometer**
As shown in figure 2-1, automobile speedometers are carefully calibrated by the manufacturer before installation. After installation, however, the speedometer's accuracy can be affected by factors such as tire size, temperature, barometric pressure, and mechanical wear as the device ages.

In most states a driver can travel up to 8 km/h (5 mi/h) over the posted speed limit without being subject to arrest for speeding. The law recognizes that some error is to be expected in a measurement made with a speedometer. Tread wear alone can contribute an error of up to 3 km/h, a variation that is well within the legal limit. But if you decide to mount a set of oversized tires on your car without recalibrating the speedometer, you are asking for a speeding citation. The point to remember is that the acceptable amount of measurement error always depends on the application. ◀

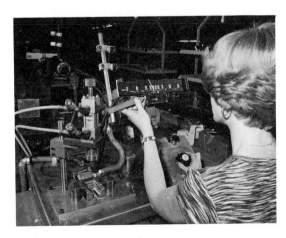

Figure 2-1. *Calibration of automobile speedometers. Regardless of the care taken at the factory, conditions of use can contribute to speedometer errors. (Courtesy of General Motors Corporation)*

"True Value"

Now let's look further at what we mean by *error*. Suppose a radio transmitter is putting out a 52-watt signal but that our instruments measure the output power level to be 47 watts. According to our definition, the measurement error here is 5 watts. It seems simple enough, but let's stop and think a little. If we knew ahead of time that the "true value" was 52 watts, we would have had no reason to make the measurement that gave us the 47 watts.

We make measurements only when we want to find out something we do not know already. On the other hand, if our instruments told us that the power output was 47 watts, we would have had no way of knowing that the "true value" was 52 watts. We can generalize a bit. Since we can never know the value of any physical quantity unless we measure it, and since no measurement is absolutely accurate, it follows that *we can never know the "true value" of any physical quantity*. This is why "true value" is enclosed in quotation marks. When we make a measurement, we usually assume that there *is* such a thing as a true value, yet at the same time we recognize that we will never know exactly what this "true value" is.

This uncertainty severely limits the usefulness of our definition of *measurement error*, for it means we can never calculate exactly what our measurement error is. The best we can do is estimate how big the *probable error* is.

Probable Error

▶ **Example 2-2: Electrical resistors**
Since electrical resistors are mass produced, manufacturers usually cannot afford to measure the resistance of each individual resistor they produce. In a batch of several thousand, only a small number will actually be measured, then the entire batch will be coded with the average measured value. Yet because actual resistance is bound to vary slightly from one resistor to another, a buyer has to know how accurate the coded value is likely to be.

To help judge, the manufacturer encodes the probable error range on the resistor. Sometimes it will be printed as a numerical percent (usually for large-wattage resistors), and sometimes it will be included as part of a color code. In any case, the buyer is told that the "true value" resistance may differ from the labeled value by up to 1 percent, or 5 percent, or 10 percent, or even twenty percent. If the buyer needs the "true value" to be very close to the labeled value, he or she will have to pay more for a resistor with a lower percentage of *probable error*. ◀

▶ **Example 2-3: Resistor coding**
A certain resistor is coded 470 ohms, 10%. What are the upper and lower limits of the "true value" resistance?

Ten percent of 470 ohms is 47 ohms, which means that the "true value" may be as high as 470 ohms plus 47 ohms, or 517 ohms. Or it might be as low as 470 ohms minus 47 ohms, or 423 ohms. The manufacturer has therefore certified that the actual resistance is somewhere between 423 ohms and 517 ohms.

What if we measured the resistance *very accurately* and found it to fall outside this range? We would have to conclude one of two things: either there is a mistake in the resistor's label or else the resistor itself is unreliable. Either way, we should probably discard the resistor. ◄

Note that the last example clearly stated that our own measurement was very accurate. If it had not been, we would have had to attribute part of the discrepancy to our own measurement error.

Repetition

In most measurement work, we will be in a position like that of the resistor manufacturer. Time allows us to make only relatively few measurements, yet we have to make some very general statements about our results. And people who need to use our data will certainly want to know something about the probable errors in the figures we give them.

As a starting point, we have to recognize that repetition is a key to reliability in measurements. In fact, it is impossible to estimate the probable measurement error if we make only a single reading. A line surge can throw off a single voltage measurement; a defective sample can give a nonrepresentative result in a tensile test; even a simple measurement with a steel rule can be off because of a human mistake in reading the scale. These things happen often enough that we must *expect* them to happen every once in a while. Thus make it a practice to *never report a measurement that is the result of only a single trial.* Throughout this book, we will always assume that each measurement has been repeated several times at least.

Systematic and Random Errors

We can now make a distinction between two basic types of measurement errors: We need to make this distinction because these errors are handled in different ways.

DEFINITION | *A systematic error remains the same throughout a set of measurement trials.*

> DEFINITION | A *random error* varies from trial to trial and is equally likely to be positive or negative.

Sources of Systematic Error

Of the two types, systematic errors are usually the more difficult to detect and account for. Systematic errors generally originate in one of two ways:

1. **Errors of calibration.** If the measuring instrument is not brought into precise agreement with a standard, or if the standard itself is not a faithful reproduction of a primary standard, then all readings from the instrument will be affected in the same way, giving rise to a systematic error. For instance, any measurement of a time interval on a clock that gains time will be too large.

2. **Errors of use.** If the instrument is not used under conditions identical to those prevailing when it was calibrated, that change of conditions may affect the way the instrument responds to the quantity being measured. Again, all the measurements in a set of trials will be affected in the same way, and the error is systematic. For instance, if a steel tape measure was calibrated at a temperature of 20°C but is being used at a temperature of -10°C, thermal contraction causes all the measurements to come out slightly too high.

Dealing with Systematic Error

If the only errors in a set of measurements are systematic, the numerical results themselves will give no clue that an error exists. Why? Because if each measurement in a set of trials is thrown off by an identical amount, all the numbers come out the same. We should constantly keep in mind that close agreement in the results of several trials does not eliminate the possibility of a systematic error.

What can be done to minimize systematic errors? First, it's important to fully understand the instrument and the physics of its operation. We should know how the instrument's accuracy is likely to be affected by temperature, humidity, and barometric pressure. We should know exactly how to calibrate the instrument, and how often the instrument usually needs to be recalibrated. If the instrument was last calibrated under conditions different from those that currently prevail, we may have to perform a recalibration on the spot. If a recalibration is not practical, we may have to correct our readings mathematically.

Since the range of instruments is so enormous, it is often necessary to evaluate systematic errors on many instruments we have never encountered in formal training. To do so, we have to rely on the manufacturer's operating manuals. We also need to have a reasonable understanding of basic physics, or else we can never judge which factors are likely to affect the equipment's operation.

Detecting a Systematic Error

▶ **Example 2-4: The thermocouple thermometer**

High temperatures may be measured by using a very sensitive voltmeter to read the electromotive force (emf) generated by an iron-constantan thermocouple (figure 2-2). The thermocouple has two junctions; one is maintained at a standard reference temperature of $0.0°C$, the temperature of melting ice, and the other is raised to the unknown temperature. The voltmeter reading depends on the difference between the two temperatures. (Remember, a measurement always involves a comparison.)

If the hot junction is at $211.0°C$, for instance, the voltmeter will read 11.39 millivolts (mV). At $212.0°C$, the reading will go up to 11.45 mV. Suppose that the temperature must be known to within 1 C°. It follows then that the voltmeter must be accurately calibrated to within 0.06 mV— by no means a simple task. Yet even with the voltmeter accurately cali-

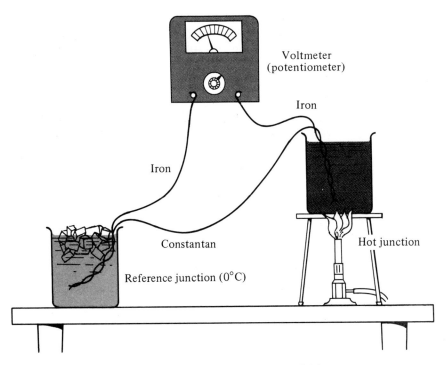

Figure 2-2. *Using a thermocouple to measure a high temperature.*

brated, there still could be a systematic error of more than 1 C°. Only a very methodical technician with some physics knowledge would recognize this.

Where would this additional systematic error come from? From the ice water. True, ice always melts at the same temperature it freezes at: 0.0°C. Barometric pressure affects this temperature by no more than a few ten-thousandths of one degree. But if the ice was made from tap water, which contains dissolved inorganic salts, the freezing–melting temperature could easily be lowered by 1 C° or more. (This is the principle behind using salt on icy streets.) Thus the technician who is using the thermocouple to make a very accurate measurement should be extra careful that the ice was made from pure, *distilled* water. ◄

Once we know that a systematic error exists in a measurement, we can often figure how to eliminate it, or at least how to make it small enough to neglect. Of course, discovering a systematic error is not always easy, so it is wise to be on guard against them constantly.

Sources of Random Error

Random errors, however, are another story. Although their origin may also be very subtle, we at least have ways of dealing with them mathematically. In most measurements, only random errors will contribute to estimates of probable error.

Random errors arise because of either uncontrolled variables or specimen variations.

1. **Uncontrolled variables.** These variables are minor fluctuations in environmental or operating conditions that cause the instrument to respond differently from one measurement trial to the next.
2. **Specimen variations.** If the measurement trials are being made on a number of presumably "identical" samples, minor differences in chemistry, physical structure, optical properties, etc., between one measurement specimen and another, will give rise to random errors.

Dealing with Random Error

Suppose that we are measuring the boiling point of a certain liquid. We properly calibrate our instruments, then make a series of temperature readings. As shown in table 2-1, the measurements actually vary from one trial to the next despite all our care and accuracy.

Does this mean that our measurements are no good? Not at all. All it

Table 2-1. *Results of a Boiling-Point Measurement Affected by Small Fluctuations in Uncontrolled Variables*

Trial	Boiling Point (°C)
1	317.51
2	317.72
3	317.22
4	317.93
5	317.02
6	317.83

means is that there are minor uncontrolled variables at work. These uncontrolled variables contribute to the random errors we see in the data. Vibrations, air current, fluctuations in the line voltage, friction in the meter movement—all these will have some minor effect on the measurement. Yet as a practical matter, these variables are very difficult to control. If we do not need any more accuracy than about 0.5°C, it is pointless for us to worry about them.

Suppose, however, that the measurements yielded the results shown in table 2-2. Here again we see the work of an uncontrolled variable, but this time it is causing some very large random errors. If we are after accuracy, this set of data is obviously of little use. Our only choice is to trace down the major cause of these errors, then to take practical steps to keep the variable responsible under control. Perhaps there is a defective component in one of the instruments, or perhaps the thermocouple has a bad connection. If the readings are taken in an enclosed container, perhaps the pressure is varying wildly. Obviously, the specific remedy must depend on which of these (or other) factors is causing the problem. In any case, the idea is to control all variables that can influence the result of the measurement, and to control them closely enough that the resulting random errors are no longer objectionable.

How can random errors result from specimen variation? We already dis-

Table 2-2. *Results of a Boiling-Point Measurement That Is Affected by Sizable Variations in Uncontrolled Variables*

Trial	Boiling Point (°C)
1	302.13
2	365.08
3	324.19
4	349.98
5	303.71

cussed error in resistor manufacture, so let's take that example further. Electrical resistance is a quantity that can easily be measured very accurately, but it is much more difficult to keep the variables in the manufacturing process under close control. As a result, there is a fair amount of variation from one mass-produced resistor to another, even in the same batch. If the manufacturer quotes a single average value for all the resistors in the batch, then the resistance errors in the individual resistors will vary in a random way. The random error here is rooted in the specimen rather than in the measuring instrument.

A similar situation is encountered when the measurement is destructive. One example is a measurement to determine the ultimate shear strength of a rivet. As always, the measurement must be repeated several times to be reliable. The problem is that the first measurement trial involves destroying the rivet. Thus the second trial must be made on a different rivet, which is also destroyed. If six trials are made, six different rivets have to be used. Because uncontrolled variables are always present to some degree in any manufacturing process, these six rivets will not be identical. We expect, then, that each rivet will fail at a slightly different value of shearing stress. When we quote a value for the undestroyed rivets, we have to realize that the "true value" actually varies from specimen to specimen. In other words, random errors come from the minor variations in the rivets' physical and chemical composition. These errors are built into the rivets themselves, and no amount of painstaking care in the measuring process can do anything about them.

It is very important to understand these two different ways that random errors can arise. Some measurements, of course, will exhibit random errors that arise from both sources. But if the major contribution comes from specimen variation, it may be useless to waste time and money on precise control of variables during the measurement itself. When the errors are already built into the system, the only thing to do is to report that they are there, and to estimate how large they are likely to be.

Summary

Because comparisons with standards are always subject to practical limitations, no measurement can ever be exact. The acceptable error in a measurement will depend on the reason the measurement is made. Since a quantity's "true value" is never known exactly, the error will never be known exactly either, and it becomes necessary to estimate the *probable error*. The probable error in a measurement is reduced by repeating the measurement several times.

Errors may be classified as *systematic* or *random*. Systematic errors are rooted in the instrument's manufacture, in its calibration, or in the conditions of its use. Random errors arise from uncontrolled variables or specimen variation. The methods for estimating probable error in the next chapter will be based on the distinction between these types of errors.

REVIEW QUESTIONS

1. What is a measurement error?
2. What is the difference between an error and a mistake?
3. What determines how large an error is acceptable in a measurement?
4. Why isn't it possible to know a physical quantity's "true value"?
5. Why is it necessary to quote values of the probable error rather than the actual error?
6. Why should you never report a measurement that is the result of a single trial?
7. What is a systematic error?
8. What are the two ways that systematic errors may originate?
9. What is a random error?
10. What are the two ways that random errors may originate?
11. What is an uncontrolled variable? Give an example.
12. What is meant by specimen variation? Give an example.
13. Give an example of a destructive measurement.
14. Why is it important to know where errors originate?

EXERCISES

1. Suppose you are measuring the pressure in your car's tires. How much measurement error would you consider allowable? (*Note:* There is no single right answer to a question like this; your answer will depend on how you drive your car, how you load it, how much you value your tread life, and how much you value your personal safety.)

2. Suppose you hang an outdoor thermometer outside your bedroom window so you know how to dress when you rise in the morning. How much measurement error would you consider allowable here?

3. Setting a spark plug gap is a process that involves a trial setting followed by a measurement, followed by an adjustment, followed by another measurement, and so on. Suppose that the manufacturer's data indicates that 0.001 in. of electrode is burned away for each 1 000 mi of driving.
   a. How much measurement error would you consider allowable in your setting?
   b. What factor or factors contributed to your answer?

4. A certain electrical resistor is coded 5 230 ohms, 1%. What are the upper and lower limits on the "true value" resistance?
(*Answer:* 5 178 ohms and 5 282 ohms)

5. An electrical resistor is coded 330 ohms, 10%. What are the upper and lower limits on the "true value" resistance?

6. Suppose that a properly calibrated thermometer is hung outside a window to indicate outside temperature. What factor or factors might contribute to a systematic error favoring high readings?

7. A simple way of comparing the aerodynamic drag on two cars is as follows: Take one car onto a level, lightly traveled road, accelerate to 50 miles per hour, throw the transmission into neutral, and have the passenger time how long it takes the car to decelerate to 40 miles per hour. Repeat the procedure for the other car. The shorter the deceleration time, the larger the aerodynamic drag. (*Note:* It is actually possible to get a numerical value for the drag coefficient by a simple extension of this procedure.)
   a. Based on your experiences in driving and riding in cars, list all the variables you think might contribute to random errors in such a measurement.
   b. What steps might be taken to bring some of these variables under control? Which variables might be impractical to control?
   c. What sources of systematic error might be present in this measurement?

8. A certain swimming pool is to be shaped like a guitar. To estimate the volume of water needed to fill the pool, it is first necessary to find the surface area. One way to do so is to make a very accurate scale drawing of the pool on a sheet of stiff paper, then, on the same paper and the same scale, to make an accurate drawing of, say, a square 10 feet on a side. Cut both shapes out with a razor blade, then weigh them on a precision balance. Dividing the weight of the pool-shaped piece by the square piece will give a comparison of the areas. What sources of error might be present in such a procedure?

CHAPTER 3

Dealing with Uncertainties

Error and Uncertainty

By now it should be clear that measurement errors are unavoidable. Because of them, some *uncertainty* will always surround any measurement result. In this chapter, we will develop methods for making a best estimate of the "true value" of a measured quantity, and for determining the uncertainty in that figure.

DEFINITION | <u>Uncertainty</u> *is the estimated amount by which the observed or calculated value of a quantity may differ from the "true value." It also may be called the probable error.*

The uncertainty in a measurement will be due to both systematic and random errors. Most of the time, one of these will overshadow the other. The best procedure, then, is first to determine whether the error is mostly systematic or mostly random, then to estimate how large this error is likely to be. This estimate is the *uncertainty* in the measurement.

Correction for Systematic Error

With systematic errors, we can sometimes (but not always) make a correction to the measurement based on this estimate. The following example shows one way to do so.

▶ **Example 3-1: Thermal contraction of a tape measure**

A 30-metre steel tape measure is designed for use at a temperature of 20.0°C. Suppose that we need to use the tape outdoors when the temperature is only −9.0°C. Since the steel tape will contract in the cold, we expect a systematic error to result. And because the scale divisions are getting closer together, we expect the measured results to be too high.

How much contraction is there? The handbooks tell us that steel contracts by 11 millionths of its length for each 1 C° drop in temperature. We have a total temperature drop of 29 C°, so the total fractional contraction is

$$(29)(11 \times 10^{-6}) = 319 \times 10^{-6}$$

where the notation 10^{-6} represents one one-millionth.

We can now calculate the total contraction in centimetres:

$$(319 \times 10^{-6})(30 \text{ m})\left(\frac{100 \text{ cm}}{1 \text{ m}}\right) = 0.96 \text{ cm}$$

This would be the systematic error for each 30-metre measurement if we neglected the contraction due to temperature.

If we have reason to believe that this is the only systematic error, we can easily correct our measurements by subtracting this 0.96 centimetres for each measured 30 metres. In other words, once we have taken the time to figure out how large our systematic error is, we no longer need to have the error. ◀

Correction for systematic error is an essential part of many measurement procedures. The following story is a good illustration. An engineer who was working on carburetor improvements used a dynamometer to measure an engine's power output. He then removed the carburetor, made some changes at the machine shop, then a few days later reinstalled the carburetor and repeated the power measurement. To his horror, he found that the power output had dropped by 10 horsepower. The measurement, however, was susceptible to systematic error due to changes in atmospheric conditions, mainly barometric pressure. When the engineer corrected mathematically for these effects, he found that the engine's power output had actually *increased* by some 25 horsepower. If he had not made the correction, he would have concluded that his carburetor alteration had hurt the engine's power rather than helped it.

Uncertainty and Random Error

So much for systematic errors, at least for a while. The rest of this chapter will concern uncertainties that result from *random* errors.

Table 3-1 shows a set of measurement results where a random error is evident. This particular measurement was repeated five times, and five different values of optical reflectivity resulted. What single value should be quoted for the reflectivity?

Table 3-1. *Data on a Certain Mirror's Optical Reflectivity to Light of 6243 Å Wavelength*

Trial	Reflectivity*
1	0.824
2	0.816
3	0.832
4	0.821
5	0.815

*Optical reflectivity is a unitless quantity whose maximum value is 1.

Median

If the uncontrolled variables are truly acting at random, they are just as likely to make a single trial come out too high as they are to make it come out too low. So if we had a very large number of trials, we would expect exactly half to be too high and half to be too low. If we arranged the trial results in order from high to low, the result in the exact center of the list would be a reasonable estimate of the "true value." This number is the *median* of the data.

DEFINITION | The *median* of a set of data is the value that falls in the exact middle of a list that orders the data from high to low. If two entries fall in the middle, the *median* is taken as the average of the two.

Finding the Median

▶ Example 3-2: The median of the data in table 3-1
We begin by arranging the data from high to low values:

$$0.832$$
$$0.824$$
$$0.821$$
$$0.816$$
$$0.815$$

The center entry here is 0.821. Exactly two entries are higher, and the same number are lower. The value 0.821 is therefore the median. ◀

Arithmetic Mean or Average

Although the median is occasionally used as the best estimate of the "true value," the *arithmetic mean*, or *average*, is more commonly used.

> **DEFINITION:** The <u>arithmetic mean</u>, or <u>average</u>, of a set of data is the algebraic sum of all the data values divided by the total number of trials. If the measurement is made to establish the value of some quantity x, we represent the arithmetic mean of the data by $\bar{x}$ (read this as "x bar").

Finding the Mean

▶ **Example 3-3: The arithmetic mean of the data in table 3-1**
Adding the five values gives a total of 4.108. We then divide this total by the number of trials, 5, to get

$$\bar{x} = \frac{4.108}{5} = 0.822$$

Thus the arithmetic mean, or average, is 0.822. ◀

Agreement between Median and Mean

Notice that the median and the arithmetic mean for this data—0.821 and 0.822—are in very close agreement. With this kind of agreement, it is of little concern which we use. For the sake of consistency, the arithmetic mean will be used in this book for the best estimate of a measurement result.

Why, then, was the median even mentioned? Two reasons. First, you should be familiar with the term just in case you ever encounter it in your work. But second, and most important, the median gives a check on the mean. Under normal circumstances, with truly random errors, the two should be in fairly close agreement. A large discrepancy between the two may indicate that something is wrong with the data or the way it has been analyzed.

Take, for instance, the set of data in table 3-2. This measurement is made to determine the heating value of natural gas. The measurement has units of Btu (British thermal units) per cubic foot; in other words, the quantity of heat per volume of gas. The difficulty is that the sixth trial gives a value much larger

Table 3-2. *Data on the Heat of Combustion of Natural Gas*

Trial Number	Heat of Combustion (Btu/ft^3)
1	999
2	1 004
3	1 008
4	1 019
5	1 011
6	1 038
7	991
8	1 014
9	1 001
10	996

than all the other trials do. From this, we can legitimately question whether this trial is valid. Perhaps someone misread an instrument, or perhaps the gas pressure or the air pressure fluctuated significantly. On the other hand, maybe it is a perfectly good data value. How can we decide?

Although it is impossible to be sure without making a large number of additional trials, we can make a reasonable judgment by looking at the mean and the median. As mentioned before, the mean and median should be fairly close together if the errors are random. If throwing out the suspect trial brings the mean and median closer together, the oddball data value was probably influenced by factors other than those producing the random fluctuations in the rest of the data. In that case, we are justified in discarding that data value. The next example shows the procedure.

Discarding a Data Point

▶ **Example 3-4: Best value for the data in table 3-2**
We begin by arranging the data in order from highest to lowest:

$$\begin{array}{c} 1\ 038 \\ 1\ 019 \\ 1\ 014 \\ 1\ 011 \\ 1\ 008 \\ 1\ 004 \end{array}$$

$$1\ 001$$
$$999$$
$$996$$
$$991$$

The median, according to our definition, is midway between the fifth and sixth entries. Thus

$$\text{median of ten trials} = 1\ 006\ \text{Btu/ft}^3$$

The mean is calculated in the usual way, by adding all the data values and dividing by 10. This gives

$$\text{mean of ten trials} = 1\ 008.1\ \text{Btu/ft}^3$$

Ordinarily, this result would be our best value for the measurement. But the highest data value was suspect so let's see what happens if we ignore it. Nine trials remain, and the median is the middle value:

$$\text{median of nine trials} = 1\ 004\ \text{Btu/ft}^3$$

The mean, or average, is calculated by adding these nine remaining entries, then dividing by 9. This gives

$$\text{mean of nine trials} = 1\ 004.8\ \text{Btu/ft}^3$$

We see, then, that discarding the suspect point brings the mean and median closer together. We therefore should do just that. The best value for the result is $1\ 004.8\ \text{Btu/ft}^3$.

We are not justified in discarding more than one data point this way, unless we have a very large number of trials, say, 100 or more. Further, the discarded point must always be the one farthest away from the median of the original data set.

Let's get on with the matter of the uncertainty. We have settled on the mean as our best estimate of the "true value." Remember, we do not really *know* what the "true value" is; the mean just gives us an estimate for it. Now the question is, How do we establish the uncertainty in this estimate?

Again, we have several choices. The most commonly used quantities are the *maximum deviation,* the *mean deviation,* and the *standard deviation.* Unfortunately, these quantities will never agree with each other the way our median and our mean agreed. We have to decide on the one that best suits our purpose in a given measurement situation.

Summation Notation

Each of the methods for determining uncertainty is based on deviations from the mean. Introducing some symbols here will make things a bit easier. If we are measuring a quantity x, we have already agreed to represent the arithmetic mean by $\bar{x}$. This mean is calculated by averaging the results of the individual measurement trials. We represent the values of these trials by x_1, x_2, x_3, and so on up to the last trial. The small numbers at the lower right of the x's are called *subscripts;* they are just a way of keeping track of which trial we are talking about. A formula for calculating the arithmetic mean of five trials can now be written this way:

$$\bar{x} = \frac{1}{5}(x_1 + x_2 + x_3 + x_4 + x_5) \qquad (3\text{-}1)$$

Since this formula is a bit cumbersome, let's shorten it a bit. The symbol Σ (capital Greek sigma) will represent the *algebraic sum of what follows.* Instead of the numerical subscripts, we will use the letter i as a single subscript. The symbol x_i still represents the result of a measurement trial, but no single trial in particular. When $i = 2$, however, x_i becomes x_2 and we are talking about the result of trial 2. Then equation (3-1) can be shortened to

$$\bar{x} = \frac{1}{5} \sum_{i=1}^{5} x_i \qquad (3\text{-}2)$$

This means that the subscript i assumes all integer values from 1 to 5, that the algebraic sum is calculated for all the x's with these subscripts, and finally that this result is multiplied by 1/5 to get the mean. This, of course, is exactly what we already did in example 3-2. All we have done here is summarize the procedure in a kind of mathematical shorthand.

We will not always have exactly five trials: sometimes there may be 7, sometimes 10, sometimes even more. Let's let the letter N represent this total number of measurement trials. Then equation (3-2) becomes

$$\bar{x} = \frac{1}{N} \sum_{i=1}^{N} x_i \qquad (3\text{-}3)$$

Deviation

We can now use this same symbolic notation in discussing uncertainties.

Since each of the three methods of discussing uncertainties will involve *deviations* from the mean, let's begin by defining *deviation*.

DEFINITION | The *deviation*, or *deviation from the mean*, is the difference between a single trial and the mean of the set of trial results. In symbols,

$$d_i = x_i - \bar{x} \qquad (3\text{-}4)$$

Calculating Deviations

▶ **Example 3-5: Deviations of data in table 3-1**
We have already calculated the mean of this data to be 0.822. Since there were five trials, there will be five deviations. Subtracting the mean from each trial result gives the following:

i	x_i	d_i
1	0.824	0.002
2	0.816	−0.006
3	0.832	0.010
4	0.821	−0.001
5	0.815	−0.007

Note that this procedure is bound to give us some negative deviations. Also note that the algebraic sum of all the deviations comes out very close to zero (−0.002 in this case).* ◀

The simplest way of quoting the measurement uncertainty is to give the *maximum deviation*. Doing so is fairly straightforward, as the example 3-6 shows.

Using the Maximum Deviation

▶ **Example 3-6: Result of the measurement in table 3-1**
Let's quote the best result for the measurement trials in table 3-1, and give the uncertainty as the maximum deviation.
Looking over the deviations in the last example, we see that the maximum is 0.010. We may then give the result as

$$x = 0.822 \pm 0.010$$

What we are saying here is that the mean, 0.822, is probably very close to the "true value." But we are also saying that our measurements show that we could be wrong by as much as 0.010. True, we did not have any negative deviations quite that large, but since random errors are equally likely to be positive or negative, we settle on an uncertainty of plus or minus a single figure. ◀

The maximum deviation is commonly used in *tolerancing* applications. We have already seen one example: the tolerance figure on the resistance of a mass-

---

*It is sometimes stated that the algebraic sum of the deviations is exactly zero. This statement is true if the mean is not rounded off when it is calculated. As we will see in a later chapter, however, there are good reasons for rounding off the mean as we have done here.

produced resistor. This tolerance figure defines the *maximum deviation* that the resistance value of any single resistor can have from the stated value. If the manufacturer says the resistance is, say, 470 ohms ± 10%, that means that no individual resistor deviates from 470 ohms by more than 10%, or 47 ohms. In other words, 47 ohms is the maximum deviation.

Tolerance

In chapter 2, we saw that random errors arise from two sources: uncontrolled variables and specimen variations. If we are trying to quote a value of a single property of many identical manufactured items (resistors or rivets, for instance), specimen variation is usually the main problem. In that case, the standard practice is to indicate the uncertainty by quoting the maximum deviation. In such applications, the maximum deviation is also often called the *tolerance*.

DEFINITION | The maximum deviation, or tolerance, is the largest amount by which a single measurement can deviate from the mean.

Many measurement situations involve no specimen variation because just one thing is being measured. We might, for instance, be measuring the thrust of a prototype of a new jet engine and have only one engine. In that case, any random errors can be due only to uncontrolled variables.

Obviously, if we have only one specimen and our measurement does not destroy it in the process, we can repeat the measurement as many times as we want. We should expect, further, that the more times we do repeat the same measurement, the more our uncertainty will decrease. Unfortunately, if we use the maximum deviation as our measure of uncertainty, exactly the opposite is true: the maximum deviation is actually likely to get *larger* as we make additional measurements. In any case, there is certainly no way it can get smaller. Consequently, we seldom quote the maximum deviation in cases where we can make repeated measurements on the same specimen. One alternative is to quote the *mean deviation*.

Mean Absolute Deviation

DEFINITION | The mean deviation–or more properly, the mean absolute deviation–is the average of the absolute values of the deviations from the mean. In symbols,

$$\bar{d} = \frac{1}{N} \sum_{i=1}^{N} |x_i - \bar{x}| \qquad (3\text{-}5)$$

We have already seen that the deviations themselves can be positive or negative. When we calculate the mean deviation from equation (3-5), we must ignore these signs or else we will always get a result very close to zero. The two vertical bars in this formula—the symbol for absolute value—are telling us to do just that.

Calculating Mean Deviation

▶ **Example 3-7: The mean deviation for the measurement trials in table 3-1**
We still have $N = 5$; that is, there are five trials in the set. We already calculated the five deviations in example 3-5. Applying the formula for mean deviation, we have

$$\bar{d} = \frac{1}{N} \sum_{i=1}^{N} |x_i - \bar{x}|$$

$$\bar{d} = \frac{1}{5}(0.002 + 0.006 + 0.010 + 0.001 + 0.007)$$

$$= \frac{1}{5}(0.026)$$

$$\bar{d} = 0.005\ 2\ , \text{ or } 0.005$$

The measurement result then may be stated as

$$x = 0.822 \pm 0.005 \blacktriangleleft$$

Standard Deviation

Statisticians prefer to estimate uncertainty by calculating a quantity called the *standard deviation*.

DEFINITION | The *standard deviation of the mean*, represented by σ_m (lowercase Greek sigma), is a quantity that can be calculated from the formula

$$\sigma_m = \frac{5}{4N\sqrt{N-1}} \sum_{i=1}^{N} |x_i - \bar{x}| \qquad (3\text{-}6)$$

where N, the number of trials, is greater than or equal to 5.*

Equation (3-6), sometimes called "Peter's formula," is the simplest of several formulas for calculating the standard deviation. Peter's formula is accurate only for five or more trials. Values of the coefficient $5/(4N\sqrt{N-1})$ are listed for reference in table 3-3.

Since the data in table 3-1 is getting a little stale by now, we will not bother to work out its standard deviation. You can verify that the result is 0.003, a value that falls below the maximum deviation of 0.010 and the mean deviation of 0.005.

Table 3-4 lists data that might result from a torque measurement on a certain diesel engine. Let's assume that any systematic errors have been properly attended to, and that the application requires us to quote the best result and its uncertainty. Subject to these assumptions, the next example shows how to analyze this data.

Table 3-3. *Values of the Coefficient in Peter's Formula (equation 3-6)*

N	$\dfrac{5}{4N\sqrt{N-1}}$
5	0.125 0
6	0.093 2
7	0.072 9
8	0.059 1
9	0.049 1
10	0.041 7
15	0.022 3
20	0.014 3
25	0.010 2
50	0.003 57
100	0.001 26

---

*Although we will often hear this quantity referred to as simply the standard deviation, we should be aware that technically it is the *standard deviation of the mean*. For our purposes, which are descriptive rather than analytical, we need not worry about the distinction. In applications like statistical quality control, however, such distinctions must be made very carefully.

Table 3-4. *Data on the Torque Produced by a Certain Diesel Engine Rotating at 1 100 Revolutions per Minute*

Trial	Torque (N·m)
1	453.2
2	450.2
3	453.8
4	455.1
5	451.9
6	454.9
7	452.5

Calculating Standard Deviation

▶ **Example 3-8: Torque from a diesel engine**

Looking over the data, we see no values that look wildly inconsistent. The median is 453.2 newton-metres (N·m). Calculating the mean gives

$$\bar{x} = 453.09$$

Since the two are in very close agreement, we need not think about discarding a data point.

We will use the standard deviation to indicate the uncertainty. Calculating this quantity from equation (3-6) requires that we first compute all the absolute deviations $|x_i - \bar{x}|$:

i	x_i	$|x_i - \bar{x}|$
1	453.2	0.11
2	450.2	2.89
3	453.8	0.71
4	455.1	2.01
5	451.9	1.19
6	454.9	1.81
7	452.5	0.59

We then sum these values to get

$$\sum_{i=1}^{7} |x_i - \bar{x}| = 9.31$$

Peter's formula tells us to multiply this quantity by the factor $5/(4N\sqrt{N-1})$. For $N = 7$, we can either calculate this ourselves or read the value from table 3-3. This gives

$$\sigma_m = (0.072\ 9)(9.31)$$
$$\sigma_m = 0.679$$

The measurement result is written as

$$x = \bar{x} \pm \sigma_m$$

or in this case,

$$x = 453.09 \pm 0.679\ \text{N·m}$$

We usually, however, quote the uncertainty to no more than two-place accuracy:

$$x = 453.09 \pm 0.68\ \text{N·m} \blacktriangleleft$$

Confidence Level

The standard deviation is the statistician's favorite way of expressing uncertainty because it has a very specific statistical interpretation. There is a 68% probability that the "true value" lies in a range of $\pm \sigma_m$ about the mean. There is a 95% probability that the "true value" lies in a range of $\pm 2\sigma_m$, or two standard deviations, about the mean. And there is better than a 99% probability that the "true value" lies within three standard deviations of the mean. When we quote the standard deviation as our measurement uncertainty, anyone who uses our results knows exactly how to interpret them.

What do these probabilities mean? Much the same as a meteorologist's claim that there is, say, an 80% chance of rain on a given day. The meteorologist is saying that, given the identical meteorological conditions a great many times, he will predict correctly 80% of the time. We are saying that, given a great deal of data from which we can calculate a large number of means, 68% of these means will be within one standard deviation of the "true value." We may refer to this 68% as the *confidence level*.

DEFINITION | The <u>confidence level</u> is the probability that the "true value" lies within a specified uncertainty range about the mean.

Obviously, there will be times when a 68% confidence level is just not good enough. We might have to be 99% sure, for instance, that the heat of combustion of a certain commercial fuel lies within a given range. Yet the standard deviation automatically gives us the 68% confidence level. Is there any way we can adjust the error limits to give a different confidence level? The answer is yes. One of the conveniences of using the standard deviation is that the user of the result can easily adjust the uncertainty limits to give any confidence level he or she wants. Table 3-5 lists the information needed to do so.

Table 3-5. *Confidence Levels for the "True Value" Lying within Certain Ranges about the Mean*

Confidence Level (Percent)	Range (Number of Standard Deviations)
50	0.67
68.3	1.00
80	1.28
90	1.65
95	1.96
95.4	2.00
96	2.05
98	2.33
99	2.58
99.7	3.00

Adjusting Uncertainty According to Confidence Level

▶ **Example 3-9: Voltage surge protection**

A certain line voltage measurement is made and quoted as

$$V = 23\,540 \pm 150 \text{ V}$$

where 150 V is the standard deviation. A breaker is to be installed in the line as a precaution against surges. The breaker must be adjusted to a voltage that will not be triggered under normal conditions. If the acceptable confidence level is 99%, to what voltage level should the breaker be adjusted?

From table 3-5, we see that the 99% confidence level corresponds to 2.58 standard deviations. We know that one standard deviation is 150 V. Thus the allowable positive deviation is

$$(2.58)(150 \text{ V}) = 387 \text{ V}$$

We can round this off to 390 V (few instruments could measure these extra 3 V anyway). Then the minimum setting for the breaker becomes

$$23\,540 + 390 \text{ V} = 23\,930 \text{ V}$$

In practice, we would probably be justified in being a little conservative and rounding this figure upward to 24 000 V. ◀

Repetition and Uncertainty

One other feature of the standard deviation is that it tends to be smaller when the number of trials is larger. This is consistent with the earlier statement that repetition is the key to accuracy in measurement. The more trials that are made, the lower the uncertainty; the fewer trials that are made, the larger the uncertainty.

Does this mean that we always need to repeat our measurements a large number of times? Not necessarily; in many cases two or three spot readings are sufficient. The point is that when random errors are present and we need to quote the result formally, we are forced to make enough trials to allow us to estimate the uncertainty mathematically.

One trial does not tell us anything, since there is nothing else to compare it with. It may be close to the "true value" or it may be miles away. A second measurement trial gives some idea of whether a random error is present. The problem with just two trials, however, is that the mean of two numbers is always the same as their median, so two trials can never lead to the rejection of a data point. Thus three trials is the practical minimum, and five is the magic number if we need to attach a level of confidence to our results. We need five trials to calculate a standard deviation from Peter's formula.

Scale-limited Error

Let's now take a look at a few slightly different examples. Examples 3-10 and 3-11 represent a very common class of measurement results, in which all the trial results are identical. In such cases we say that we have *scale-limited errors*.

DEFINITION | *Scale-limited errors are said to exist when the random errors are masked by the coarseness of the measuring instrument's scale divisions.*

▶ **Example 3-10: Electric current**
The following data represents the results of a current measurement. We want to quote the best value for the current and the uncertainty in this quantity.

2.5 amperes (A)	2.5
2.5	2.5
2.5	2.5

The mean here is obviously 2.5 A. Since there is no deviation from the mean in any trial, the maximum deviation, the mean deviation, and the standard deviation are all zero. But is the uncertainty really zero? Let's be careful. The entire discussion in this chapter has been based on the existence of random errors, and there are no random errors in this data. What do we do now?

The first thing to do is to check the instrument to see what the smallest scale division is. Suppose that it is 0.1 A, and that the instrument has a certified accuracy consistent with this figure. Then we must conclude that the readings were made to the *nearest* tenth of an ampere. If any reading was more than halfway to 2.6 A (the next scale division), it would have been recorded as 2.6 A. This means that all the trials fell within ±0.05 A of 2.5 A. We therefore report the result as

$$x = 2.5 \pm 0.05 \text{ A}$$

We cannot claim that this uncertainty has any relation to the standard deviation. There is no way to attach statistical significance to it, other than to say that it represents a "fairly high" level of confidence. ◄

▶ **Example 3-11: Length**
A certain length measurement yields the following data. We need to quote the best value and its uncertainty.

3.16 m	3.16
3.16	3.16
3.15	3.16

This data looks very much like the data in the last example, except for the single low entry. Keeping the value of 3.15 m gives us a mean that slightly disagrees with the median. Rejecting the 3.15 value brings the mean and median into exact agreement. According to the rule developed earlier, the low reading should be rejected.

But now we have five identical trial results. The situation is the same as that in the last example. If we do not have the measuring instrument to check (or the measurer to talk to), we have every reason to conclude that the measurements were made to the nearest 0.005 m. Our result, then, is

$$x = 3.16 \pm 0.005 \text{ m}$$ ◄

Chaotic Error

We should be aware of one other type of error: the *chaotic error*. It lies at the other extreme from the scale-limited error.

Table 3-6. *Chaotic Errors in Data on the Optical Power Output of a Certain Laser*

Trial	Power Output (mW)
1	2.2
2	12.7
3	18.9
4	0.6
5	7.9
6	22.4
7	0.1

DEFINITION | *Chaotic errors* are said to exist when fluctuations in the data are so large that the uncertainty becomes comparable to the mean.

Table 3-6 shows a set of data where chaotic errors are evident. The spread in the data alone tells us that something is drastically wrong. If we did take the trouble to calculate the mean and the standard deviations, the result for the power output of the laser would be

$$P = 9.3 \pm 3.5 \text{ mW}$$

At a 95% confidence level, all we could say from this is that the "true value" lies somewhere between 2.4 and 16.2 mW, which is not saying very much. Obviously, the problem is either in the measuring equipment or in the laser itself, and there is no way that just taking and analyzing more data will improve the situation.

Summary

Every measurement result will have an *uncertainty*, or probable error. If the source of error is clearly apparent, a mathematical correction to the measurement can sometimes be made. More commonly, however, an uncertainty must be reported as part of the result.

Random errors are analyzed statistically, as summarized in table 3-7. The *mean*, or average, of a set of trial results is taken to be the best estimate of the "true value." If an extreme data point is suspected of being a mistake, it may be rejected if doing so will bring the mean into closer agreement with the *median*. If the random error is due primarily to specimen variation, the uncertainty is expressed as the *maximum deviation* (*tolerance*). If the random error is primarily due to uncontrolled variables, the uncertainty is expressed as the

Table 3-7. *Summary of Methods of Stating Uncertainty When Random Errors Are Present*

Cause of Fluctuations	Best Value	Uncertainty
specimen variation	mean (usually)	maximum deviation
uncontrolled variables	mean	mean deviation or standard deviation

mean absolute deviation or the *standard deviation.* The standard deviation, which establishes a confidence level for the uncertainty, should not be calculated for fewer than five trials.

Random errors may also be *scale limited* or *chaotic.* Scale-limited errors imply an uncertainty equal to one-half of the instrument's smallest scale division. Chaotic errors imply that the entire set of results should be discarded and the measurement repeated from the beginning.

We have seen that a great deal can be said about a measurement's accuracy by analyzing the data. To get a complete picture, however, we have to ask some questions about the instruments, too. We will look into this matter in the next chapter.

REVIEW QUESTIONS

1. What is measurement uncertainty?

2. Explain how it is sometimes possible to correct a result for a known systematic error.

3. What is the median of a set of data?

4. What is the mean of a set of data?

5. Identify the meaning of each symbol in the equation

$$\bar{x} = \frac{1}{N} \sum_{i=1}^{N} x_i$$

6. What is the rule for discarding a questionable entry in a set of data?

7. What is deviation, or deviation from the mean?

8. Suppose that a certain measurement is repeated seven times. How many deviations from the mean are there?

9. Why do the sum of the deviations from the mean always come out close to zero?

10. What is maximum deviation?
11. What is another term commonly used in practice to mean the same thing as maximum deviation?
12. What is mean absolute deviation?
13. Identify the meaning of each symbol in the equation

$$\bar{d} = \frac{1}{N} \sum_{i=1}^{N} |x_i - \bar{x}|$$

14. What is standard deviation?
15. What is Peter's formula used for?
16. How many measurement trials must be made for Peter's formula to be accurate?
17. What is confidence level?
18. How many standard deviations correspond to a confidence level of 50%? of 80%? of 99%?
19. Why can't we estimate the uncertainty after only one or two trials?
20. What is the preferred method of stating uncertainty when the errors are random and due to specimen variation?
21. What is the preferred method of stating uncertainty when the errors are random and due to uncontrolled variables?
22. What are scale-limited errors?
23. What are chaotic errors?

EXERCISES

1. Comparison with a standard shows that a certain mechanical clock gains 8 seconds every 250 seconds. The clock is used to measure a time interval, giving a result of $t = 7$ min 28 s. Assume that the random error here is scale limited, and correct the result for the systematic error.

2. A sonar rangefinder measures underwater distances by timing how long it takes for a sound pulse to travel to the underwater object and back to the instrument. If this time is represented by t, the distance x is $x = \frac{1}{2}vt$, where v is the underwater speed of sound. In most cases, this calculation is done within the instrument itself.

   One such instrument is calibrated for use at a water temperature of 25°C. Since the speed of sound varies with water temperature, a systematic error will result if the instrument is not used at that temperature. Determine the size of this systematic error on a measurement result of x = 267.3 m when the water temperature is 16°C. The speed of sound in sea water is 1 531 m/s when the temperature is 25°C. This figure decreases by 2.4 m/s for each 1 C° drop in temperature. (*Answer:* The result is too large by 3.8 m.)

3. The following data represents the result of a magnetic field measurement. The unit is gauss (G).

$$
\begin{array}{ccc}
2\,365 & 2\,341 & 2\,367 \\
2\,360 & 2\,353 & 2\,335 \\
2\,372 & 2\,382 & 2\,375
\end{array}
$$

Find the median and the arithmetic mean. (*Answer:* median = 2 365 G; mean = 2 361.1 G)

4. The following data represents the result of a pressure measurement. The unit is newtons per square centimetre (N/cm^2).

$$
\begin{array}{cc}
89.7 & 90.3 \\
86.2 & 92.5 \\
89.9 & 91.6 \\
88.1 & 88.7
\end{array}
$$

a. Determine whether any entries should be rejected.
b. Quote the best value of the result and its mean deviation. (*Answer:* No rejections; $x = 89.6 \pm 1.5$ N/cm^2)

5. The following data represents the result of a viscosity measurement. The unit is centipoise (cp).

$$
\begin{array}{cc}
5.273 & 5.280 \\
5.291 & 5.298 \\
5.286 & 5.278 \\
5.337 & 5.275
\end{array}
$$

a. Determine whether any entries should be rejected.
b. Quote the best value of the result and its mean deviation.

6. Find the standard deviation of the data in exercise 4. (*Answer:* 0.694 N/cm^2)

7. Find the standard deviation of the data exercise 3.

8. Find the standard deviation of the data in exercise 5.

9. Express the result of the measurement data in exercise 4:
   a. at a 50% confidence level
   b. at a 68.3% confidence level
   c. at a 99.7% confidence level
   (*Answer:* The uncertainty ranges are ± 0.47 N/cm^2, ± 0.69 N/cm^2, and ± 2.08 N/cm^2, respectively.)

10. Express the result of the measurement data in exercise 5:
    a. at an 80% confidence level
    b. at a 95.4% confidence level
    c. at a 98% confidence level

11. Express the best result of the data in table 3-2, with an uncertainty corresponding to the 90% confidence level.

CHAPTER 4

Accuracy versus Precision

Instrument Accuracy

You may have had the experience of going to a department store to buy an outdoor thermometer. On the display racks you see a number of mechanical thermometers in several models and price ranges. Each thermometer has one Celsius degree as the finest scale division. You begin to puzzle over your selection when you notice, to your horror, that each thermometer is indicating a different temperature.

Now obviously, the room can be at only one temperature. Something must be wrong with at least some of the thermometers. Your first suspicion is that the thermometers were sloppily calibrated. That certainly is a possibility, but it probably is not the whole answer. There is another problem here, a subtle one at best.

A thermometer is calibrated by making it coincide with at least two of the fixed points of the IPTS (table 1-2). For the normal temperatures we are dealing with, the points used would be number 5, the standard boiling point of water, and number 6, the triple point of water. (Since the triple point is only 0.01 C° from the freezing point of water, it would be just as appropriate and a great deal easier in this case to use the freezing point as the second standard.) Now the problem is this: Even if all the thermometers were to agree at the boiling and freezing points, we have no automatic guarantee that they will agree at every temperature in between. In fact, an experienced metrologist (a scientist who studies the methods and problems of measurement) would be very surprised if two thermometers *did* agree at every temperature.

We have already seen that errors in the manufacturing process prevent the

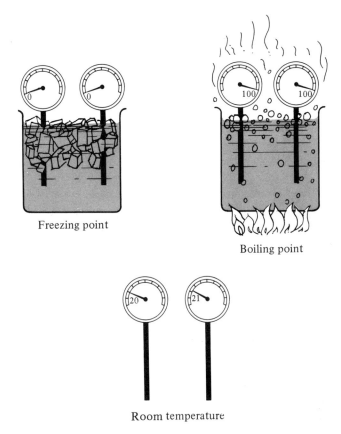

Figure 4-1. *Agreement of instruments at a calibration point does not guarantee their agreement at some other point. Two properly calibrated thermometers can easily disagree at room temperature.*

production of a batch of perfectly identical parts. That means that the two thermometers in figure 4-1 are not identical. As a result, they will not respond to changes in temperature in identical ways. So even though a thermometer is accurately calibrated, its accuracy is still suspect at points other than the calibration points. The scale may be divided into degrees, or even fractions of a degree, and yet the accuracy might be no better than plus or minus a few degrees.

We will come back to thermometers in chapter 12. Right now, let's concentrate on what this word *accuracy* means.

DEFINITION | *The accuracy of an instrument is a certification of how closely it can be expected to agree with its calibration standard.*

When we hear the word *accuracy,* we should immediately think in terms of physical standards and instrument calibration. Yet we should remember that an

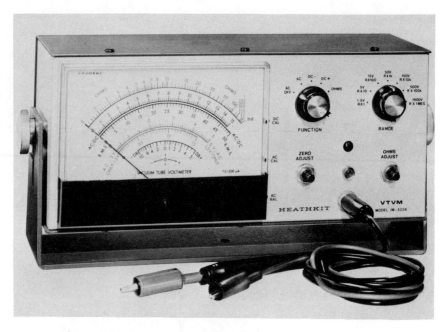

Figure 4-2. *Specifications for a Heath IM-28 multimeter. Notice that the instrument's accuracy as a dc voltmeter is ±3% of full-scale, while as an ac voltmeter its accuracy is ±5% of full-scale. (Courtesy of Heath Company)*

IM/SM-5228 SPECIFICATIONS: DC VOLTMETER—Ranges: 0–1.5, 5, 15, 50, 150, 500, 1500 V full scale; up to 30,000 V with accessory probe. *Input Resistance:* 11 megohm (1 megohm in probe) on all ranges: 1100 megohms with accessory probe. Circuit: Balanced bridge (push-pull) using twin triode. *Accuracy:* ±3% of full scale. *AC VOLTMETER—Ranges:* 0–1.5, 5, 15, 50, 150, 500, 1500 rms scales. *Frequency Response (5 V range):* ±1 dB 25 Hz to 1 MHz (800 ohm source, referred to 60 Hz). *Accuracy:* 5% of full scale. *Input Resistance & Capacitance:* 1 megohm shunted by 40 pF measured at input terminals (200 pF at probe tip). *Ohmmeter—Ranges:* Scale with 10 ohm center, X1, X10, X100, X1000, X10k, X1 meg. Measures .1 ohm to 1000 megohms with internal battery. *Meter:* 6" 200 μA movement, polystyrene case. *Battery:* 1½ V, "C" cell (not supplied). *Power requirement:* 120/240 VAC, 60/50 Hz, 10 W. *Dimensions:* 5" H × 12 11/16" W × 4 3/4" D.

instrument's overall accuracy cannot be improved simply by improving on the calibration. The calibration points of our thermometers might agree with the IPTS fixed points as closely as 0.1 C°, yet they can still disagree in midscale by several degrees. If that is the case, we will gain nothing by trying to improve on the calibration. The ultimate accuracy of an instrument is determined in the manufacturing process, and we can do very little to improve it after the fact.

The manufacturer of any quality instrument will report its accuracy in the operating manual or specifications sheet, so it is important to be acquainted with this information before using the instrument. Figure 4-2 shows a manufacturer's specifications for a multimeter.

Accuracy of a Meter Scale

▶ **Example 4-1: Voltmeter accuracy**
Electrical instruments often use the same scale for several measurement ranges. A voltmeter scale, for instance, may look like this:

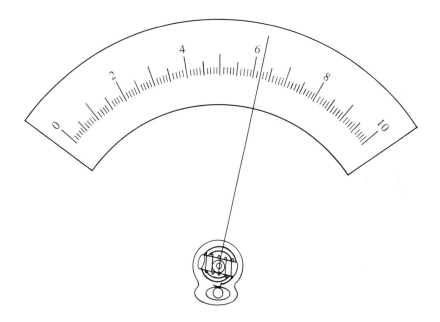

This particular scale is divided into 100 parts. The largest number (full-scale) is 10, but this does not mean that the voltmeter is reading 10 V when the pointer is on 10. To get the actual reading, we have to look at the position of the range switch, which might look like this:

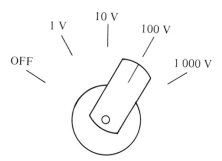

This range switch is telling us that a full-scale deflection requires 100 V. This means that the number 10 on the scale really represents 100 V; the number 2 is really 20 V, and the smallest scale division is 1 V. If the range switch is changed to the 1-V position, the smallest scale division represents 0.01 V.

Because this voltmeter is a multirange instrument, the manufacturer cannot specify the accuracy as a blanket range of plus or minus a certain number of volts. Instead, the accuracy will be given as, say, ±3% of full-scale deflection.

Suppose that the needle indicates 6.2 when the range switch is in the 100-V position. What is the result of the measurement? First, we notice that the 10 on the scale is really 100 V. This means that the 6 indicates 60 V, and 6.2 becomes 62 V. Since the full-scale deflection is 100 V and the accuracy is ±3% of full-scale, this measurement is accurate to within ±3 V. The result, then, is

$$V = 62 \pm 3 \text{ V}$$

Notice that the measurement error may be as large as 3 V even though the smallest scale division represents 1 V. ◄

Accuracy and Systematic Error

From the preceding example, we see that the instrument's accuracy directly affects the measurement error. Note that this ±3 V is due to systematic error in the measurement. This systematic error exists even though the voltmeter is properly calibrated. It comes about because the meter is calibrated at only one or possibly two voltages, while it is used to give a reading at other voltages between 0 and 1 000 V where it was not calibrated. If we wanted to, we could trace this error back to errors in the manufacturing process. That, however, would not make it go away. It is there, and it is built into the instrument.

It is quite possible for an instrument to contribute a systematic error to a measurement, while at the same time uncontrolled variables are producing a random error. In such cases, we need to consider both sources of error when we write the measurement uncertainty. The next example shows how this is done.

Combined Systematic and Random Error

▶ **Example 4-2: Voltage measurement**

A certain voltage measurement is made on the 10-V scale of a meter with an accuracy of ±2% of full-scale. The measurement is repeated five times, giving the following values:

7.1 V
7.0
7.1

7.2
7.3

What is the best value result and its uncertainty?

First, let's find the probable error due to the meter itself:

$$(\pm 0.02)(10 \text{ V}) = \pm 0.2 \text{ V}$$

This is the range of the possible systematic error in *each* of the five measurement trials. We do not know the actual systematic error (if we did, we could correct for it), but we do know that it might be as high as 0.2 V. If it is, then each of the five trials will be off by this same 0.2 V. This means that we will have to add 0.2 V to the probable random error in the data.

The average of this data is 7.14 . This gives a mean deviation of 0.09. Therefore the uncertainty due to random error might be given as ±0.09 V. Adding the uncertainty due to systematic error, we get a total uncertainty of ±0.29 V. The final result is

$$V = 7.14 \pm 0.29 \text{ V} \blacktriangleleft$$

Voltmeter Applications

After all this discussion about voltmeters, you may wonder if you should really be concerned about voltmeters if you are not working in electronics. The answer is yes, you should. First, the principles illustrated in these last two examples apply equally well to pneumatically operated and mechanical instruments. But second, voltmeters are commonly used to measure many things other than voltage. Any energy form that can be converted to a voltage can be measured with a voltmeter. Strain gauges, light meters, pH meters, thermocouple thermometers, certain types of vacuum and pressure gauges, radiation-monitoring instruments, and many other measuring devices use some form of a voltmeter. And the principle is the same whether the voltmeter has a moving-coil movement (as in the preceding examples) or whether it has a digital readout.

Instrument and Measurement Accuracy

Our definition of the word *accuracy* states that it is a property of the instrument. Occasionally, however, the word will be used as if it is a property of the measurement. This usage should cause no confusion if you remember that an accurate measurement is one that is made with a properly calibrated, accurate instrument.

We have seen that an instrument's accuracy may not be as good as its finest scale division. The voltmeter in example 4-1, for instance, could be read to the nearest volt, yet the accuracy was only within 3 V. That gave our 62-V reading a built-in uncertainty of ±3 V. (Remember that this uncertainty means that the actual voltage might have been anywhere between 59 and 65 V.)

Why, then, did the manufacturer even bother to put 1-V divisions on the scale? Are they really good for anything? Yes, the 1-V divisions can be used to *discriminate* between two voltages as close as 1 V apart. Example 4-3 shows how.

Instrument Discrimination

▶ **Example 4-3: Voltage comparison**
The voltmeter in example 4-1 is used to determine how temperature affects a battery's voltage. First the voltage is measured when the battery is at 15°C. Let's call this result V_1. The battery is then warmed up to 20°C, and its voltage is measured again. Let's call this second result V_2.

Suppose that random errors in each measurement are less than 0.5 V, and that the two results are

$$V_1 = 24 \text{ V}$$
$$V_2 = 26 \text{ V}$$

What can we say about the effect of temperature on the battery?

Notice that the problem is not to find the actual voltage, but rather to find how much the voltage *changes*. And since the meter scale can discriminate down to 1 V, we can be confident here that the voltage has increased by about 2 V. This is our result: an increase in temperature from 15°C to 20°C increases the battery voltage by about 2 V.

What is the actual voltage in each case? That is another question. Based on the ±3-V accuracy, all we can say about V_1 is that it is between 21 and 27 V. As for V_2, it is somewhere between 23 and 29 V.

So although the meter can give an actual voltage only to within 3 V, it can give a voltage *change* to within 1 V—the meter's smallest scale division. ◀

Precision

An instrument's accuracy, therefore, does not give the whole story of its usefulness. If the instrument is used to measure small changes in a quantity, and it is these *changes* that are important, then we need to know how small a change the instrument can detect. This value is called the instrument's *precision*.

DEFINITION | The precision of an instrument is the index of its discriminating ability. It is usually stated as the instrument's smallest scale division.

A common ruler, for instance, may divide an inch into 16 parts. If so, we can say that the ruler's precision is $\frac{1}{16}$ of an inch. Such a ruler can be used to discriminate between two lengths, provided that they differ by at least $\frac{1}{16}$ inch.

Precision and Accuracy

As we have seen, it is possible for an instrument to have a precision that exceeds its accuracy. A thermometer may have a precision of 0.2 C° but an accuracy of ±1 C°. This means that it can give the actual temperature only within 1 C°, but that it can register small temperature changes of as little as 0.2 C°.

Is the situation ever reversed? Will we ever encounter instruments whose accuracy exceeds their precision? Only rarely, because it makes little sense to manufacture such an instrument. Suppose that a voltmeter has a precision of 1 V and an accuracy of ±0.1 V. The meter will always be in very good agreement with its calibration standard; in fact, it will never be off by more than 0.1 V. The problem is that we cannot *read* the instrument this finely. We can read it only to its scale precision—the nearest volt. All that extra accuracy is wasted.

It is a great deal more expensive to build accuracy into an instrument than it is to build precision into it. So if the instrument is engineered to be very accurate, it will also be engineered to be very precise. This is simply a matter of economics. We pay for the accuracy, but without precision the accuracy becomes useless. Ergo, the manufacturer might as well give us the precision, too. The thing to remember is that many instruments are more precise than they are accurate, whereas very few instruments are a great deal more accurate than they are precise.

In some instruments—particularly in many of the common dimensioning instruments—accuracy and precision run hand in hand. A steel tape, for instance, may have divisions of 1 mm. That, then, is its precision. Its accuracy is probably ±0.5 mm. If so, then the tape accuracy is consistent with its precision.

Let's see what these figures mean. We make a measurement with the rule, and reading to the nearest scale division we get a result of

$$x = 113 \text{ mm}$$

What do we quote as the uncertainty? If we know that the accuracy of the rule is ±0.5 mm or better, we can say that we have a scale-limited error. In such cases, standard practice is to give the uncertainty as one-half of the precision. The result is

$$x = 113 \pm 0.5 \text{ mm}$$

Suppose instead that the rule has an accuracy of ±0.2 mm but the same precision of 1 mm. Our measurement can still be off by as much as half the smallest scale division, so the improved accuracy is of no benefit. The result is still

$$x = 113 \pm 0.5 \text{ mm}$$

Now suppose that the precision is still 1 mm but that the accuracy is only ±1 mm. We can read to the nearest scale division, but that division itself may be off by ±1 mm. In this case, it is standard practice to assume that the entire systematic error is determined by the accuracy of the rule itself. We therefore quote the result as

$$x = 113 \pm 1 \text{ mm}$$

Table 4-1. *Precision of Common Dimensioning Instruments*

Ordinary Ruler	Steel Rule
1/16 inch	0.02 inch
1 millimetre	0.5 millimetre
Steel Tape	*Vernier Caliper*
0.1 inch	0.001 inch (±0.002-inch accuracy)
1 millimetre	0.1 millimetre
Standard Micrometer Caliper	*Vernier Micrometer Caliper*
0.001 inch	0.0001 inch (±0.000 3-inch accuracy)
0.01 millimetre	0.001 mm (±0.005-mm accuracy)

Note: Unless otherwise indicated, the instrument's accuracy is plus-or-minus one-half of its stated precision.

As mentioned, many instruments will have an accuracy consistent with their precision. Table 4-1 lists the precisions of 6 common dimensioning instruments. All these instruments are available in both English and metric versions, so 12 instruments are actually listed. Of these 12, all but 3 have an accuracy consistent with their precision. The English vernier caliper, however, has a precision that exceeds its accuracy, so both figures are listed. The same is true of the English and metric vernier micrometer calipers. In fact, the metric vernier micrometer caliper has a precision that so far exceeds its accuracy that the instrument offers almost no advantage over the standard metric micrometer caliper. As a result, it is a rather unusual instrument to find in use.

Converting Instrument Accuracy to Uncertainty

▶ **Example 4-4: Micrometer caliper measurement**
A standard metric micrometer caliper is used to measure the diameter of a steel rod. The measurement is repeated three times, and all three readings are 3.51 mm. What is the uncertainty?

Since there are no random errors here, any uncertainty is systematic (i.e., relating to the instrument itself and how it is used). We see from table 4-1 that the accuracy of a metric micrometer caliper is consistent with its precision. This means that the uncertainty is one-half of the smallest scale division. Our result is

$$d = 3.51 \pm 0.005 \text{ mm} \blacktriangleleft$$

▶ **Example 4-5: Vernier micrometer caliper measurement**
An English vernier micrometer caliper is used to measure the diameter of a bearing. The reading is 0.361 2 in., and a doublecheck gives the same result. What is the uncertainty?

From table 4-1, we see that the accuracy of the instrument is within 0.000 3 in. Again, there are no random errors. The result is therefore

$$d = 0.361\ 2 \pm 0.000\ 3 \text{ in.} \blacktriangleleft$$

Implied Uncertainty

Cases like example 4-4 are frequent. In other words, we will often find that an uncertainty is one-half of the last recorded place value. When this happens, it is common not to write the uncertainty at all. If we follow this rule, a result of

$$R = 2.921 \pm 0.000\ 5 \text{ m}$$

becomes simply

$$R = 2.921 \text{ m}$$

Notice, however, that if the uncertainty is anything other than half of the last place value in the quoted result, it has to be written out.

This process, of course, is reversible. A quoted result of 622 mℓ should be interpreted as 622 ± 0.5 mℓ, and a quoted result of 16.20 km should be interpreted as 16.20 ± 0.005 km.

The scheme may seem simple enough, but it has one problem. Suppose we weigh something on a scale with a precision of 100 lb and get a result of

$$W = 22\ 000 \pm 50 \text{ lb}$$

We decide to follow convention and write it as simply

$$W = 22\ 000 \text{ lb}$$

How is someone else to interpret this measurement? Unfortunately, it *looks* like the measurement was made to the nearest 1 000 lb, so the result would be translated as

$$W = 22\ 000 \pm 500 \text{ lb}$$

In other words, the short cut caused us to sacrifice a great deal of precision.

One way to avoid this problem is as follows: If a number is greater than 1 and the last significant digit is a 0, indicate the significance of the 0 by placing a line over it. Our weight, then, would be written as

$$W = 22\,0\overline{0}0 \text{ lb}$$

Implying Uncertainty in a Result

▶ **Example 4-6: Volume measurement uncertainty**
The volume of a storage tank is measured, with a result of

$$V = 43\,002 \pm 50 \text{ } \ell$$

Write this in its simplest possible form.

First, we have to recognize that the 2 doesn't mean very much. We might have *tried* to measure this volume to the nearest liter, but the uncertainty came out a great deal larger. And the ±50 ℓ itself is not a precise number—it is simply an indication of the probable error range. In actuality, then, we do not change the *meaning* of the quoted result by writing

$$V = 43\,000 \pm 50 \text{ } \ell$$

Notice that our uncertainty is one-half of 100. Since there is a 0 in the hundred's place, we write

$$V = 43\,\overline{0}00 \text{ } \ell$$

The line over that 0 indicates that it is significant. The actual volume is closer to 43 000 ℓ than it is to 43 100 ℓ or to 42 900 ℓ. ◀

In nontechnical literature such as newspaper and magazine articles, measurement uncertainty is always implied rather than written. This can sometimes lead to misconceptions. In 1850, for instance, the height of Mt. Everest (tallest mountain in the world) was surveyed for the first time. The result, accurate to about 6 in., was 29 000 ft. Had this figure appeared in the newspapers, readers would have naturally assumed that the measurement had been very crude. For this reason, the surveying team deliberately reported a false result of 29 002 ft. Incidentally, when the survey was repeated in 1954, the result came out as 29 028 ft. It is unclear whether the mountain actually grew in the last century, or whether one or the other of the measurements had a larger error than estimated. Many mountain climbers still prefer the earlier figure.

Let's back up now and see how all these ideas about precision and accuracy fit in with the material in previous chapters.

The Complete Measurement Process

We make measurements so we can describe a physical thing quantitatively; in other words, so we can answer the question "How much?" If the answer to this question is important, then it is also important that we have some idea of how good our answer is. The "goodness" of the answer is described by giving the measurement uncertainty.

The difficulty, as we have seen, is that measurement uncertainty may arise in a variety of ways. To estimate this uncertainty, we need to take a careful look at these potential sources of error. And how we finally estimate the uncertainty will depend on where we finally decide the measurement error came from. If the error seems to be systematic, we estimate the uncertainty one way; if it is random, we do things another way. The estimation of uncertainty, then, is not a simple step-by-step procedure.

One way to outline a procedure that involves decisions is to use a *flowchart*. Flowcharts, which are commonly used in computer programming and industrial engineering, are useful for organizing any multistep procedure where the actual sequence of steps may depend on the outcome of some of the steps. Figure 4-3 is a flowchart of the entire procedure for making a measurement. The rectangles indicate steps that involve no decision, the diamonds indicate steps that do involve decisions, and the arrows indicate the sequence of steps. Notice that there are two possible ways to go after each decision (diamond).

Now it may seem that the measurement process is terribly complicated. But we really do not have to follow a flowchart to make a measurement. The point of figure 4-3 is simply this: Anyone who makes measurements conscientiously is following a procedure very similar to the one diagrammed in the flowchart. And if there is a problem in a measurement or a very important measurement to be made, this flowchart can be used for reference.

Summary

Because of manufacturing tolerances, an instrument would not measure perfectly even if its calibration were exact (which, of course, it cannot be). An instrument's *accuracy* is a statement of how closely it can be expected to agree with its calibration standard. An instrument's *precision* is the index of its ability to distinguish between two measurements. Many instruments are more precise than they are accurate.

An instrument's accuracy should be considered in determining the uncertainty in a measurement result. The total uncertainty depends on how the accuracy is expressed, as well as on whether or not the error is scale limited. The entire procedure for determining the best result of a measurement and its uncertainty is shown in figure 4-3.

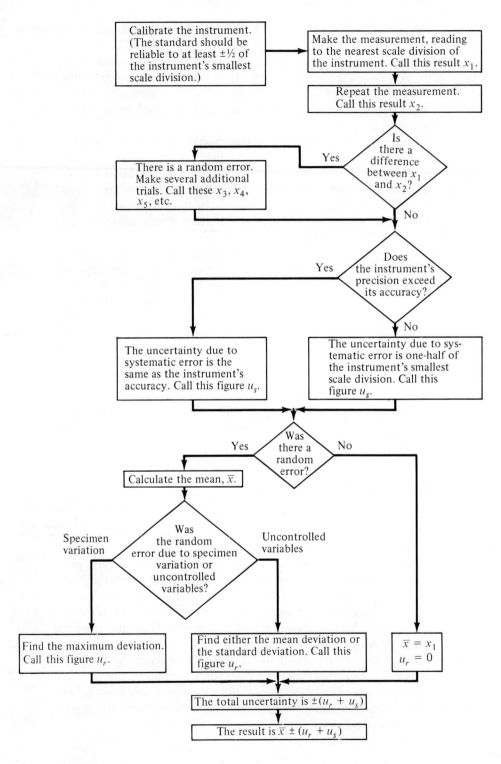

REVIEW QUESTIONS

1. What is meant by instrument accuracy?
2. How can we find an instrument's accuracy?
3. How is the accuracy of a multirange instrument specified?
4. Why can it be misleading to assume that an instrument's accuracy coincides with its smallest scale division?
5. If an error results from the limitation of instrument accuracy, will it generally be systematic or random?
6. Why are voltmeters so important in measurement applications?
7. What is discrimination?
8. What is precision?
9. What is the difference between precision and accuracy?
10. Why do we seldom encounter instruments that are more accurate than they are precise?
11. What is implied uncertainty?
12. What is a significant zero in a number? When may it lead to difficulties in interpretation?
13. What is a flowchart? What is it used for?
14. Summarize the important decisions involved in making a measurement and quoting the result.

◄ **Figure 4-3.** *Flowchart for determining the best value and the uncertainty in a measurement.*

EXERCISES

1. A voltage measurement is made with the voltmeter's range switch on the 1 000-V position. The voltmeter's accuracy is 5% of full-scale. The scale looks like the diagram in example 4-1, and the pointer indicates 4.3. Random error is negligible. What is the result of the measurement and its uncertainty?

2. Do exercise 1 for a voltmeter accuracy of 2% of full-scale. (*Answer:* 430 ± 20 V)

3. Do exercise 1 for the range switch on the 0.1-V position.

4. An electric current is measured on the 1.0-mA scale of a meter with an accuracy of 1% of full-scale. Five trials are made, with the results shown below. What is the best value result and the total uncertainty? (Use the mean deviation.)

$$0.213 \text{ mA}$$
$$0.221$$
$$0.209$$
$$0.225$$
$$0.205$$

5. A certain pressure gauge has an accuracy of ±2.0 millimetres of mercury (mm Hg). A pressure measurement is made, with the six trial results shown below. Quote the best value result, and the total uncertainty. (Use the mean deviation.)

$$117 \text{ mm Hg}$$
$$120$$
$$115$$
$$101$$
$$122$$
$$116$$

6. Express the total uncertainty in exercise 5 at the 95% confidence level. (*Answer:* 118.0 ± 6.4 mm Hg)

CHAPTER 5

Data Tables

A single measurement answers a single question. For example: What is the voltage of this battery? Or, what is the diameter of this ball bearing? The measurement may involve more than one trial, but if it answers only one question, we consider it to be one measurement. We have already seen that the result of such a single measurement should be reported in the form

$$x = \text{(best value)} \pm \text{(uncertainty)}$$

Questions Requiring More Than One Measurement

We also may encounter questions that require us to make *more* than one measurement: How is the power output of an engine related to its angular speed, for instance? How does water temperature affect the underwater speed of sound? Or, how does pressure affect the boiling point of an antifreeze liquid? To answer questions like these, we need to make a whole series of measurements. A data table is one way of presenting answers to questions like these.

Let's begin by considering the case of the wind chill index. A warm object placed outside will cool at a rate that depends on the outside temperature. But if the wind is blowing, the cooling rate is increased. One way of describing this increased cooling rate is to say that the effective temperature with the wind blowing is lower than the temperature indicated by a thermometer. This effective temperature is what we refer to as the wind chill index. Obviously, the wind chill index depends on both the actual temperature and the wind speed.

Suppose we want to find out how the wind chill index is related to temperature and wind speed. So we start making measurements. If the temperature is 0°F and the wind is blowing at 10 mi/h, we find that a warm object cools at

Table 5-1. *Wind Chill Index (in Effective °F) for Various Combinations of Actual Temperature and Wind Speed*

		Actual Temperature (°F)										
	Calm	35	30	25	20	15	10	5	0	-5	-10	-15	-20
Wind Speed (mi/h)	5	33	27	21	16	12	7	1	-6	-11	-15	-20	-26
	10	21	16	9	2	-2	-9	-15	-22	-27	-31	-38	-45
	15	16	11	1	-6	-11	-18	-25	-33	-40	-45	-51	-60
	20	12	3	-4	-9	-17	-24	-32	-40	-46	-52	-60	-68
	25	7	0	-7	-15	-22	-29	-37	-45	-52	-58	-67	-75
	30	5	-2	-11	-18	-26	-33	-41	-49	-56	-63	-70	-78
	35	3	-4	-13	-20	-27	-35	-43	-52	-60	-67	-72	-83
	40	1	-4	-15	-22	-29	-36	-45	-54	-62	-69	-76	-87

Source: National Oceanic and Atmospheric Administration

the same rate as if the wind were not blowing at all and the temperature were -22°F. We say, then, that the wind chill index is -22°F even though the actual temperature is 0°F. When the temperature is 15°F and the wind is blowing at 20 mi/h, our measurements show that the cooling effect is the same as having no wind and a temperature of -17°F. If the temperature is -5°F and the wind is blowing at . . . but now things are getting complicated. We obviously need a better way of reporting results like these.

One possibility is to organize the results in a table like table 5-1. Here we see listed, in a rather small space, the results of 96 separate measurements. Notice, however, that you need not read through all 96 to find the one you want. You simply scan the column headings to find the one closest to the temperature you are interested in, scan the row headings to find a number close to the wind speed you want, then look at the intersection of the column and row to get the single measurement corresponding to that combination. If, for instance, the temperature is 5°F and the wind speed is 20 mi/h, the wind chill index is quickly seen to be -32°F.

Most of us are already used to using tables, and this whole idea may seem simple enough. Notice, however, that the table is easy to use only because a great deal of thought went into laying it out. And that is what we are interested in here: how to lay out a table so it is easy to use.

A few rules always apply. Even though they may seem like common sense, they are still sometimes ignored—with total confusion as a result.

Organization of Tabular Data

1. Label the columns with a column box heading. The column box heading in table 5-1 is "Actual Temperature (°F)."

2. Label the rows with a row box heading. The row box heading in table 5-1 is "Wind Speed (mi/h)."
3. Head each column with its own column heading. If the column headings are to be numerical, as in table 5-1, make sure they are arranged in either ascending or descending order, and that the numerical values are equally spaced. For example, the temperature column headings in table 5-1 decrease from left to right, and they decrease in consistent jumps of 5 F°.
4. Repeat rule 3 for the row headings. Reading top to bottom in table 5-1, we see that the wind speed increases in consistent jumps of 5 mi/h.
5. Enter the measurement results in their appropriate positions. Instead of writing out each uncertainty, these results are usually rounded off so the uncertainty can be interpreted as ±½ of the last significant digit.
6. In a table heading (usually above the table), indicate the measured quantity that the table is listing.

Rules 1, 2, and 6 are obvious enough. Readers will not be able to use the table unless they know what the table is listing and what the headings represent.

Rules 4 and 5 are needed to make the table easy to read. But they also indicate that *we should be thinking about the table before we even begin making the measurements*. For instance, if we had measured wind chill at a temperature of 22°F and a wind speed of 12 mi/h, the result would not be much help in drawing up this table. If the table is going to list the temperature in 5-degree jumps, then *these* are the temperatures where the measurements should be made. The same is true of the 5-mi/h jumps in wind speed. The implication, then, is that this table must be based on measurements made under carefully controlled conditions, where the variables can be made to assume the values we have decided to list in the column and row headings.

Usefulness

You may think this sounds like putting the wagon in front of the horse—deciding first what the table will look like, then making the appropriate measurements. But the sequence is not really so strange if we remember that *the only reason we make measurements at all is that we expect them to be useful*. If the usefulness of a series of measurements depends on their being presented in a table, then we have to think about this *before* we begin measuring. If we plunge into the measurements without doing so, *then* we have put the wagon in front of the horse.

What about rule 5? After all we have done to find ways to estimate the measurement uncertainty, we do not list the uncertainty at all here. Why? Again, it's a matter of convenience and usefulness: if the table is too cluttered, it becomes difficult to read. Thus we sacrifice a little bit of accuracy in favor of convenience. Let's look at an example of how this might be done.

Rounding Off Uncertainty

▶ **Example 5-1: Uncertainty in temperature**
A certain temperature measurement gives
$$T = 21.2°C \pm 0.3°C$$
Write this without stating the uncertainty explicitly.

Our result says that the actual temperature might have been as high as 21.5°C or as low as 20.9°C. Of course, our uncertainty itself is only an estimate, so these limits are not exact. We notice, however, that if we round off these upper and lower limits to the nearest degree, we get the same result: 21°C. We can claim, then, that the original result is nearly the same as saying that
$$T = 21°C$$
which the user will interpret as
$$T = 21°C \pm 0.5°C$$

Notice that we are free to make some small adjustments in the uncertainty for convenience, since this number does not represent a mathematically exact figure anyway. We cannot, however, claim that the result here is 21.0°C. That would imply that the uncertainty was only ±0.05°C, which it certainly was not. ◀

Unequal Uncertainties in Tables

What if all the measurements cannot be rounded off to the same accuracy? Can we have a table where some of the measurements are listed to a greater accuracy than others? The answer is yes. Table 5-2 is such a table; it is also an example of a table with only one independent variable rather than two.

In table 5-2, some of the densities (aluminum, copper, carbon, and iron) are listed with an implied uncertainty of ±0.005 g/cm^3. For concrete, rubber, and carbon steel, there are larger uncertainties because of sample variation. In these cases the uncertainty (taken again to be one-half of the last significant place value) must be read as ±0.05 g/cm^3. For materials like glass and tool steel, the sample variation is so great that the table lists the actual range of values. It

Table 5-2. *Densities of Some Solids at Room Temperature* (20°C)

Substance	Density (g/cm^3)
Aluminum	2.70
Carbon (graphite)	2.25
Concrete	2.4
Copper	8.93
Glass	2.4–2.8
Iron	7.86
Lead	11.3
Rubber	1.1
Silver	10.5
Steel, carbon	7.8
Steel, tool	7.8–8.7
Wood, white pine	0.37

would be senseless and misleading to try to express all the entries to the same accuracy.

Functional Tables

There is, of course, another major difference between table 5-1 and table 5-2: table 5-1 is a *functional* table, whereas table 5-2 is a *statistical* table.

DEFINITION | A *functional table* lists the value of a quantity in terms of one or two other quantities that it depends on.

In a functional table, we can expect the column and row headings to be numerical. The row headings represent different values of the independent variable described by the row box heading [in table 5-1, "Wind Speed (mi/h)"]. If there is only one independent variable, the table will have only one column. Table 5-1, however, has two independent variables, and there are column headings to indicate the second variable's values (the actual temperature in Fahrenheit degrees). The table entries are values of the *dependent* variable: the wind chill index. This quantity is called a dependent variable because its values depend on the values of the independent variables. In the mathematical sense, we may say that our dependent variable is a *function* of the independent variables, which is why we say the table is "functional." The table itself can be considered to define a mathematical function.

Statistical Tables

The other type of data table is a *statistical table.*

DEFINITION | A *statistical table* has one set of headings that is descriptive or qualitative rather than numerical.

There is no simple quantitative way to describe different substances like glass, rubber, and copper in table 5-2, nor is there any reason we would want to. This table does not represent any mathematical function or relationship; rather, it is simply a listing of comparative values of density. Tables in almanacs commonly list things like U.S. census reports, record temperatures by city, the lengths of the world's major rivers, eruption records of volcanoes, and so on. All these tables are statistical tables. They do not imply any functional relationship between the numbers listed, but they do allow us to compare, say, two rivers to see which is longer.

Now because a functional table represents a mathematical relationship, we can subject the table's entries to certain mathematical operations that would make no sense with the statistical tables. Two of these operations, *interpolation* and *extrapolation,* are important for our concerns. We will discuss interpolation first.

Interpolation

DEFINITION | *Interpolation* is the process of estimating a function's intermediate values from the tabulated values in a functional table.

Table 5-3 lists values of the acceleration due to gravity for different latitudes on the earth's surface at sea level.* Since this table shows how the gravitational acceleration depends on specific values of latitude, it is a functional table. Notice that there is only one independent variable. The two right-hand columns simply give the same quantity, the acceleration due to gravity, in different systems of units.

The latitudes in table 5-3 are given in increments (or jumps) of 5°. What if we want the acceleration due to gravity at a latitude that isn't listed? Can we figure it out somehow?

Yes, we can figure it out *approximately;* that is what interpolation is all about. We could follow any number of interpolation procedures, but here we

---

*Latitude is the measure of a point's angular distance from the earth's equator, with the vertex of the angle at the earth's geometrical center. The equator is designated as 0° latitude, with the poles at 90° north latitude and 90° south latitude.

Table 5-3. *Acceleration Due to Gravity at Various Latitudes at Sea Level*

Latitude	Acceleration Due to Gravity	
	(m/s^2)	(ft/s^2)
0	9.780 39	32.087 8
5	9.780 78	32.089 1
10	9.781 95	32.092 9
15	9.783 84	32.099 1
20	9.786 41	32.107 6
25	9.789 60	32.118 0
30	9.793 29	32.130 2
35	9.797 37	32.143 5
40	9.801 71	32.157 8
45	9.806 21	32.172 5
50	9.810 71	32.187 3
55	9.815 07	32.201 6
60	9.819 18	32.215 1
65	9.822 88	32.227 2
70	9.826 08	32.237 7
75	9.828 68	32.246 3
80	9.830 59	32.252 5
85	9.831 78	32.256 4
90	9.832 17	32.257 7

Source: U.S. Coast and Geodetic Survey

will deal only with the simplest: *linear interpolation.* Example 5-2 explains the process.

Linear Interpolation from a Table

▶ **Example 5-2: The acceleration due to gravity**
Based on the data in table 5-3, estimate the acceleration due to gravity at a latitude of 42°.

We reason as follows: A latitude of 42° is 2/5 of the way between 40° and 45°, both of which are listed. At 40°, the acceleration due to gravity is 9.801 71 m/s^2. At 45°, the figure is 9.806 21 m/s^2. The value we are after, then, is approximately 2/5 of the way between these two figures. The difference between the figures is

$$9.806\ 21\ \text{m/s}^2 - 9.801\ 71\ \text{m/s}^2 = 0.004\ 50\ \text{m/s}^2$$

and

$$C = \frac{5}{9}(F - 32)$$

This same relationship can be described through a functional table such as table 5-4. If we make an interpolation from this table, the result will be just as accurate as the data actually listed in the table. Example 5-4 shows how this is done.

Interpolation as an Alternative to Direct Calculation

▶ **Example 5-3: Fahrenheit–Celsius conversion**
Using the data in table 5-4, find the Celsius temperature corresponding to 141.00°F.
The portion of the table we are dealing with looks like this:

°C	°F
50.00	122.00
100.00	212.00

$$\frac{2}{5} \times 0.004\ 50\ \text{m/s}^2 = 0.001\ 80\ \text{m/s}^2$$

Adding this to the lower figure (the one corresponding to 40°), we get a value of 9.803 51 m/s².
The result of our interpolation, then, is that at a latitude of 42°, the acceleration due to gravity is 9.803 51 m/s².
Note this result is not quite as accurate as the data it was derived from. In fact, the detailed figures of the U.S. Coast and Geodetic Survey give the accepted value at 42° latitude as 9.803 50 m/s². ◀

Using the interpolation process, we can estimate any intermediate value that is not actually listed in a functional table. As we have just seen, this estimate will not be as accurate as the data itself. To compound matters, there is no way to judge how much uncertainty surrounds the interpolated result. Interpolation can be very useful for making estimates, but if we need extreme accuracy and/or an estimate of the probable error, we are forced into making another measurement.
Like many rules, this one has an exception. If the functional table defines a *linear* relationship between the variables, then any linear interpolation made from the data will be just as accurate as the data itself.
For instance, the relationship between Fahrenheit and Celsius temperatures can be described algebraically through the following linear equation:

Table 5-4. *Fahrenheit–Celsius Equivalents*

°C	°F	°C	°F
−273.16	−459.72	550.00	1 022.00
−250.00	−418.00	600.00	1 112.00
−200.00	−328.00	650.00	1 202.00
−150.00	−238.00	700.00	1 292.00
−100.00	−148.00	750.00	1 382.00
−50.00	−58.00	800.00	1 472.00
−40.00	−40.00	850.00	1 562.00
0.00	32.00	900.00	1 652.00
50.00	122.00	950.00	1 742.00
100.00	212.00	1 000.00	1 832.00
150.00	302.00	1 500.00	2 732.00
200.00	392.00	2 000.00	3 632.00
250.00	482.00	2 500.00	4 532.00
300.00	572.00	3 000.00	5 432.00
350.00	662.00	3 500.00	6 332.00
400.00	752.00	4 000.00	7 232.00
450.00	842.00	4 500.00	8 132.00
500.00	932.00	5 000.00	9 032.00

The difference between the listed Fahrenheit temperatures is 90.00 F°. The difference between 141.00°F and 122.00°F is 19.00 F°. The temperature we are looking for, then, is 19/90 of the way between the listed temperatures.

Taking 19/90 of 50.00 C° (the difference between the listed Celsius temperatures), we get 10.56 C°. Adding this to 50.00°C (the temperature corresponding to 122.00°F) gives the result of the interpolation: 141°F = 60.56°C. ◄

Extrapolation

Suppose we need the value of a quantity that lies beyond the range of the data listed in a table. For example, we might be interested in the wind chill index at, say, 30 mi/h and −25°F. We can project the data to values beyond those listed in table 5-1 by a process known as *extrapolation*.

DEFINITION | *Extrapolation is the process of estimating the value of a variable outside its tabulated or observed range.*

One way of extrapolating is to use the same approach as in linear interpolation. Example 5-4 demonstrates such an approach.

Linear Extrapolation from a Table

▶ **Example 5-4: Wind chill index**
Estimate the wind chill index at a temperature of $-30°F$ and a wind speed of 20 mi/h.

Looking at table 5-1, we see that the windspeed of 20 mi/h is listed, but the temperatures are listed only as low as $-20°F$. So we are looking for a temperature that is 10 F° lower than the last column in the table.

In going from $-15°F$ to $-20°F$, with the wind speed constant at 20 mi/h, we see that the wind chill index drops from $-60°F$ to $-68°F$. This is a drop of 8 F° in wind chill when the actual temperature drops by 5 F°. If the temperature were to drop another 5 F°, we would expect the wind chill to drop another 8 F°. Likewise, if the temperature were to drop 10 F°, to $-30°F$, we would expect the wind chill to drop 16 F°, to $-84°F$.

The result of the extrapolation, then, is that the estimated wind chill index at $-30°F$ and 20 mi/h is an effective $-84°F$.

This result, unfortunately, is little more than an educated guess. The actual figure given by the National Oceanic and Atmospheric Administration for these conditions is $-81°F$. ◀

In general, an extrapolation is only a rough estimate based on insufficient data. We should therefore never expect an extrapolation to be very accurate. Further, if an extrapolation represents a projection far beyond the actual data available, its accuracy will be considerably worse than an extrapolation that goes slightly beyond the data. Weather forecasts, for instance, are basically extrapolations based on measured data. And, as we all well know, long-range forecasts are considerably less accurate than short-range ones. Extrapolations are always poor substitutes for actual measurement, so they should be used only when there is no alternative.

Summary

A data table is a convenient way of organizing a large amount of data in a small space. Tables may be *statistical* or *functional,* depending on whether the row headings are expressed qualitatively or quantitatively. In either case, the table should be designed with the purpose of making the data easy to refer to and use.

If the table is functional, *interpolations* and *extrapolations* may be made from it. These procedures allow the user to obtain data values that were not actually measured and listed in the table.

If time is going to be spent making a detailed set of measurements, then certainly some time should be invested in thinking about how to organize the results. In the next chapter, we will look into a popular alternative to the table.

REVIEW QUESTIONS

1. Give an example of a question that can be answered only by making more than one measurement.
2. What is the column box heading in table 5-3?
3. What is the row box heading in table 5-3?
4. Why should numerical column headings in a table be arranged in either ascending or descending (as opposed to random) order?
5. Why is it a good idea to have numerical column headings represent equally spaced values of the column variable, as in table 5-1?
6. Why are measurement uncertainties usually rounded off in a data table?
7. What is a functional table?
8. What is a statistical table?
9. What is interpolation?
10. What type of table (functional or statistical) lends itself to interpolation?
11. What is extrapolation? How is it different from interpolation?
12. Which is generally more accurate, interpolation or extrapolation?

EXERCISES

1. Your daily newspaper probably lists the high and low temperatures measured at representative cities across the country. Pick 10 of these cities, and calculate the daily temperature variation for each one. Do this for three different days, then arrange all of your results in tabular form.
2. Many people keep their old utility bills on file. If you have done so, make up a table showing your monthly expenditures in each category—say, for fuel, electricity, telephone—for a 12-month period.
3. The torque output of an engine is related to its power output by the relation

$$T = 5\,252 \frac{P}{f}$$

where T is the torque in pound-feet (lb·ft), P is the power in horsepower (hp), and f is the frequency of rotation in revolutions per minute (r/min). Make up a functional table that lists the torque output of an engine for different values of P and f. List the power in increments of 20 hp starting at 0 hp and going up to 200 hp. List the frequency in increments of 500 r/min starting at 500 r/min and going up to 5 500 r/min. (*Hint:* Study table 5-1 carefully.)

4. Make up a functional table listing the distance (in miles) you would travel by maintaining a certain speed for a certain time. List the time in increments of 5 min, starting at 5 min and going up to 1 h. List the speed in increments of 5 mi/h, starting at 20 mi/h and going up to 70 mi/h.

5. Do exercise 4 using metric units. The distance will be in kilometres; the speed will be in kilometres per hour, starting at 20 km/h and going up to 120 km/h in increments of 10 km/h.

6. Make up a statistical table comparing the makes of cars driven by students and faculty at your institution. Count the number of cars of each different make (e.g., Buick, Ford, Volkswagen) in the student parking lot, then in the faculty parking lot. Express the number of each make as a percent of the total in that lot. Your table will have two columns—one for faculty, the other for students. It will have as many rows as you find different makes of cars.

7. Based on the data in table 5-3, find the acceleration due to gravity at 33° north latitude.

8. Using table 5-4, find the Fahrenheit temperature corresponding to 21.37°C.

9. Using table 5-4, find the Celsius temperature corresponding to 10 000°F.

10. Interpolating from the table you drew up in exercise 3, find the torque corresponding to 96.000 hp and 1 000.0 r/min. Then do a direct calculation and compare your results (*Answer:* Interpolated, 504.16 ft·lb; direct, 504.19 ft·lb)

11. Interpolating from the table in exercise 3, find the torque corresponding to 160 hp and 1 220 r/min. Then do a direct calculation and compare your results.

12. Interpolating from the table you constructed in exercise 4, find the distance corresponding to
    a. 12.3 min at 55.0 mi/h.
    b. 40.0 min at 46.7 mi/h.
    c. 37 min, 23 s at 50.0 mi/h.
    [*Answer:* (c) 31.153 mi]

13. Extrapolating from the table you constructed in exercise 5, find the distances corresponding to the following. Compare your extrapolations with direct calculations.
    a. 50 min at 150 km/h.
    b. 83 min at 110 km/h.
    [*Answer:* (a) 124.958 km extrapolated; 125.0 km calculated]

CHAPTER 6

Constructing Graphs

The Graph

We have seen that a large amount of data can be organized in a small space by using a data table. Often, however, we will want to go farther and represent the data in visual form, in a *graph*.

DEFINITION | A *graph* is a diagram that uses lines, circles, bars, or some other geometrical form to represent tabulated data.

Notice that a graph does essentially the same thing that a table does: it compresses a large amount of information into a small space. In fact, to draw a graph we usually need to have a table first. Why, then, do we bother with graphs at all? There are a number of reasons:

Advantages of Graphs

1. Graphs get the reader's attention. If it's true that a picture is worth a thousand words, then a graph is worth a thousand numbers.
2. Graphs usually permit easier reference to the data than a table does.
3. It is easier to compare one graph with another than it is to compare one table with another.
4. Graphs reveal certain features in the data that a table may not make immediately apparent. Some of these features are:
   a. Maxima: the largest data values
   b. Minima: the smallest data values

       c. Periodicity: whether the data tends to repeat itself at a definite interval
       d. Variations in slope: points where the data values change most abruptly, or least abruptly, or possibly where the change is from increasing slope to decreasing slope
5. Graphs sometimes can be drawn directly by the measuring equipment, eliminating the need for individual measurements and tabulation.

Graphs and Precision

Graphs do, unfortunately, have one major disadvantage: they often fail to display the full precision of the measurements. If one measurement result is 2.123 cm and another is 2.131 cm, the two will probably appear to be the same when they are plotted on a graph. The only way to avoid this loss of precision is to draw a very big graph, which is certainly not always practical.

Unfortunately, this fact of life is often used as an excuse for drawing sloppy graphs. In fact, it is no excuse at all. The inherent loss of precision comes from the width of the pencil or pen lines used to draw the graph. When the thickness of these lines is comparable, on the scale of the graph, to the distance representing the last significant digit of the data, this part of the data does not show up. Precision is lost, not because the graph is incorrect, but because the dots and lines themselves are too coarse.

In the last chapter, we saw that there are two types of data tables, statistical and functional. When a graph is drawn from a statistical table, it is usually a bar graph or a circle graph. Let's see how these graphs may be drawn.

Constructing a Bar Graph

▶ **Example 6-1: Composition of the earth's crust on a bar graph**
The earth's crust is composed of 92 different chemical elements. The most common of these are listed in table 6-1 (a statistical table). Represent this data on a bar graph.

In a bar graph, we indicate the abundance of each of these elements by the length of a bar. Before we draw the bars, however, we have to make up a grid that indicates the percents. We might try to draw a grid line for each 1 percent, but this would make the graph very crowded. A better approach is to draw a vertical line for each 5 percent. Representing the data by a series of bars drawn on such a grid produces the graph shown in figure 6-1.

It is important to note two things about this graph. First, it contains all the labels needed to read it. In other words, if someone turns directly to

Table 6-1. *Composition of the Earth's Crust, by Weight*

Chemical Element	Weight (%)
oxygen	46.60
silicon	27.72
aluminum	8.13
iron	5.00
calcium	3.63
sodium	2.83
potassium	2.59
magnesium	2.09

Note: The earth's crust is the solid surface of the earth only, not including the atmosphere or the oceans.

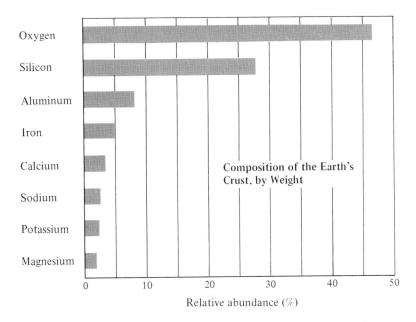

Figure 6-1. *The data of table 6-1 represented on a bar graph.*

this page and spots this graph, he or she can read it and understand it without having to hunt through the text to see what is being represented. Second, the graph cannot be read to the precision of the entries in the table. This is the point mentioned before: Graphs usually do involve some loss of precision. ◄

Constructing a Circle Graph

▶ **Example 6-2: Composition of the earth's crust on a circle graph**
Drawing a circle graph requires converting the percents to angular degrees. If we use a circle (360°) to represent 100 percent of the earth's crust, then the angle of a slice corresponding to oxygen is 46.60% of 360°, or 167.76°. We need to repeat this calculation for each of the elements in table 6-1. To keep all these calculations organized, it helps to make up a *working table*. Such a working table might look like this:

Chemical Element	Weight (%)	Angular Degrees
oxygen	46.60	167.76
silicon	27.72	99.79
aluminum	8.13	29.27
iron	5.00	18.00
calcium	3.63	13.07
sodium	2.83	10.19
potassium	2.59	9.32
magnesium	2.09	7.52
all others	1.41	5.08
totals	100.00	360.00

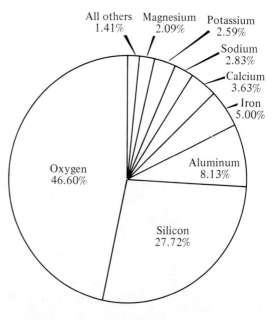

Figure 6-2. *The data of table 6-1 represented on a circle graph.*

Composition of the Earth's Crust, by Weight

Notice that we had to *add* an entry, "all others," because the original data adds up to only 98.59%. We have to presume that the leftover 1.41% corresponds to the total of all the chemical elements that are not listed.

Drawing the circle graph amounts to dividing a circle into pie-shaped wedges having the angles we calculated. This, of course, is done with a protractor. The result is shown in figure 6-2. ◀

Functional Graphs

In science and technology, we actually encounter relatively few bar and circle graphs because most of the graphs we deal with are based on functional tables rather than on statistical tables. A graph based on a functional table is called, obviously, a *functional graph*.

DEFINITION | A *functional graph* uses a line on a grid to display the relationship between two physical variables.

Figure 6-3 is a functional graph that shows the actual average temperature at any altitude between sea level and 300 km. (A commercial jetliner cruises at an altitude of about 11 km.)

We see from this graph that the temperature actually decreases for only the first 20 or so kilometres, reaching a minimum of about -80°C. After that the temperature increases for awhile, coming close to sea level temperature at an altitude of 50 km. Between 50 km and 80 km, the temperature begins to decrease again, leveling off at about -100°C. At altitudes higher than about 85 km, the temperature consistently increases with increasing altitude. In this graph it is possible to take in this *behavior* of temperature with altitude in one glance, something we could not do if the data were presented in a table. The graph's disadvantage, of course, is that we cannot read the numbers with any great precision.

Dependent and Independent Variables

Figure 6-3 displays the relationship between two physical variables: temperature and altitude. It is no accident that temperature is on the vertical axis and altitude is on the horizontal axis. In this example, altitude is the *independent variable* and temperature is the *dependent variable*. The dependent variable is always plotted on the vertical axis.

The distinction between dependent and independent variables is not as complicated as it may sound. We can send a balloon or a rocket to any altitude we choose. If we decide to measure the temperature there, we find that the

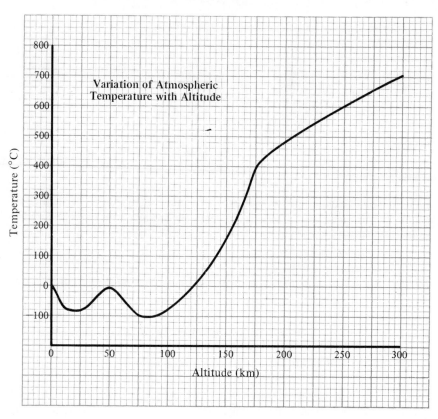

Figure 6-3. *A functional graph.*

temperature *depends* on the altitude we have chosen. Temperature, then, is the *dependent variable*. The thing it depends on—the altitude, in this case—is the *independent variable*. Notice that we cannot say that the altitude depends on the temperature. In fact, figure 6-3 shows that a temperature of -50°C can actually occur at four different altitudes at the same time.

What if we measure the aerodynamic drag of a wing at different airspeeds? It's obvious that the drag we measure depends on the speed we have chosen. The airspeed, then, is the independent variable and the drag is the dependent variable. On a graph, we would plot the airspeed on the horizontal axis and the drag on the vertical axis.

The following example shows the procedure for drawing a functional graph.

Constructing a Functional Graph

▶ **Example 6-3: Thermocouple calibration curve**

Measurements are made on the voltage output of an iron-constantan thermocouple at different temperatures. In all cases the reference junction is kept at 0.00°C. Represent the following results on a graph:

Temperature (°C)	Voltage (mV)
0.0	0.00
20.0	1.02
50.0	2.58
100.0	5.27
150.0	8.00
200.0	10.78
250.0	13.56
300.0	16.33

First we need a piece of graph paper. Suppose we use one that has 45 divisions horizontally and 60 divisions vertically. (Of course, we can turn it on its side and the opposite will be the case.)

Since our independent variable here is the temperature, we will put it on the horizontal axis. The range of this variable is 300 C°. We need to represent this range in *fewer* than 45 divisions, because we need to save some space for labels. A convenient choice is 30 divisions, because

$$\frac{300 \text{ C}°}{30 \text{ divisions}} = 10 \text{ C}° \text{ per division}$$

This horizontal scale will be both easy to plot and easy to read.

The range on the dependent variable is 16.33 mV. This has to be represented in fewer than 60 divisions. If we allow two divisions to represent 1 mV, then 32.66 divisions are needed, which leaves some remaining paper. The alternative is to allow three divisions to represent 1 mV, but working in thirds is somewhat inconvenient. So let's stay with the first choice.

The next step is to lay out and label the axes (figure 6-4a). Notice that we include only enough numerical labels to make the graph easy to use. Notice also that the numerical values are evenly spaced. In other words, if one division represents 10 C°, then each additional division should represent an additional 10 C°.

Only now do we plot the data. Each pair of measurements gives us one point on the graph, so we have a total of eight points. The last step is to connect the points with a line. The final result is shown in figure 6-4b. ◄

Interpolation in Graphs

The curve in figure 6-4b came out to be almost a straight line; in fact, we would have been justified in using a straightedge to draw it. Drawing this line amounts to filling in points that follow the same pattern as the points resulting from the

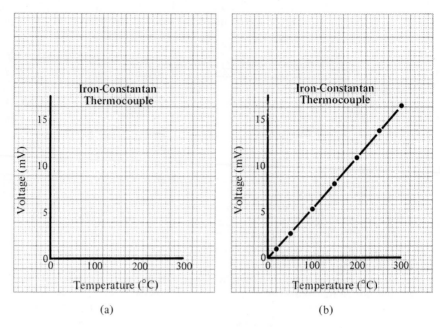

Figure 6-4. *The steps in drawing a graph. Careful attention should be paid to choosing an appropriate scale and to proper labeling.*

measurements themselves. Since all the original data points were on a straight line, we had to expect that any additional measurements would have given us points that also fell on this same straight line. In other words, drawing the graph here amounted to making a *linear interpolation* from the data.

It may seem strange to see the word *interpolation* used in connection with graphs since we originally discussed it in relation to tabular data. But when we draw a graph, we include not only our tabular data but also estimates of intermediate values that the table does not include. The process of estimating these intermediate values is exactly the same as the interpolation process we used with tables in the last chapter.

In drawing a graph, we do not actually go through the mathematical procedure for interpolating. Instead, we make the interpolation directly on our drawing. If we are doing a linear interpolation, we simply connect the data points with a single line drawn along a straightedge.

We can also make *nonlinear* interpolations on graphs by connecting the data points with curves rather than straight lines. In fact, this is the only acceptable way to treat data points that do not themselves lie on a straight line. Whereas nonlinear interpolation is a common way to prepare graphs, it is rarer in table preparation because of mathematical complications.

Nonlinear Interpolation

▶ **Example 6-4: Boiling point curve for water**
Water is known to boil at different temperatures depending on the pressure. Data is taken on the boiling point, in degrees Celsius, at various pressures. The results are shown below, where the pressures are given in *torr* (standard atmospheric pressure is 760 torr). Graph this data.

Boiling Point (°C)	Pressure (torr)
1.4	5.1
21.8	19.6
33.0	37.7
54.2	114
79.1	342
92.7	582
100	760
113	1 187
126	1 795

The independent variable, pressure, will be graphed on the horizontal axis, and the boiling point will be on the vertical axis because it is the dependent variable.

Our first problem is to choose the scale. Suppose we use a piece of graph paper with 41 horizontal divisions and 54 vertical divisions. On the vertical axis, we have 126 C° to be represented in fewer than 54 divisions. Since

$$\frac{126 \text{ C}°}{54 \text{ divisions}} = 2.33 \text{ C}° \text{ per division}$$

and we need to save some space for labels, we might decide to use a vertical scale of 1 division = 2.5C°. Notice that this is the same as saying that each 4 divisions represent 10 C°. This scale will be easy to work with, and the graph will come close to filling the paper while still allowing some extra space for the labels.

On the horizontal axis, we have 1 795 torr to be represented in fewer than 41 divisions. This gives

$$\frac{1\ 795 \text{ torr}}{41 \text{ divisions}} = 43.78 \text{ torr per division}$$

A reasonable scale to use, then, would be 1 division = 50 torr. Again, this scale leaves some space for labels.

With the scales chosen, we can sketch our axes, label them, and plot the nine data points. Finally, we have to *interpolate* between these points. We do so by drawing a single smooth curve that intersects each of them. The result is shown in figure 6-5. ◀

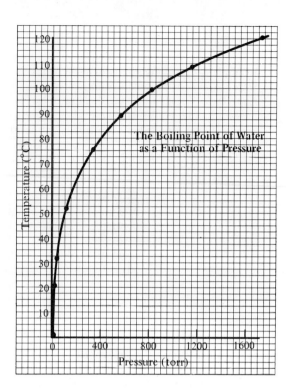

Figure 6-5. *A graph requiring nonlinear interpolation.*

Using a French Curve

When data points lie along a straight line, it is easy to interpolate by drawing a line along a straight edge. But when the points lie on a curve, as in figure 6-5, a drawing template known as a *french curve* is used. French curves are manufactured in many sizes and shapes.

Although skill in using a french curve comes only with practice, figure 6-6 shows the basic idea. First, move the french curve around until you find a section that coincides with at least three data points. Holding the french curve in this position, draw in the center part of the region of coincidence. Next, find a portion of the french curve that will extend the line through the next data point, without forming a cusp or "kink" in the line. Then, extend the line and repeat the process. The final result should be a smooth curve. Figure 6-7 shows examples of right and wrong graphical interpolations drawn with a french curve.

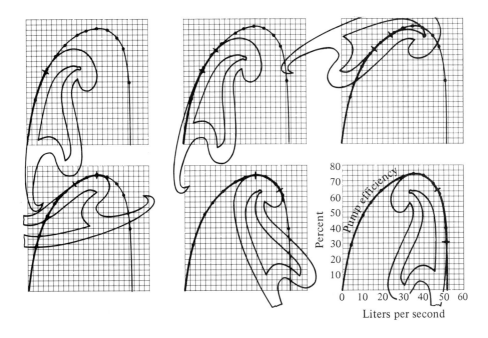

Figure 6-6. *Using a french curve to interpolate between data points not lying along a straight line.*

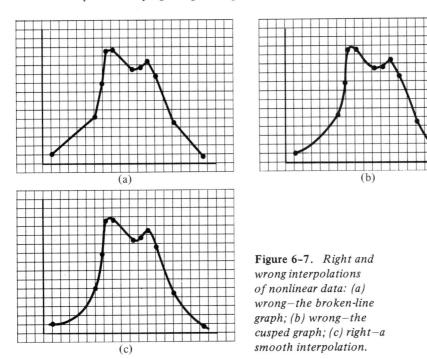

Figure 6-7. *Right and wrong interpolations of nonlinear data: (a) wrong—the broken-line graph; (b) wrong—the cusped graph; (c) right—a smooth interpolation.*

Constructing Graphs / 83

Graphing Uncertainties

So far in this chapter, we have not worried about the accuracy of our data. Of course, you should realize by now that a measurement always involves some degree of uncertainty. If an uncertainty is large enough, it must be accounted for in graphing the data.

Figure 6-8 shows how to graph an uncertainty. Here we see the results of an experiment in which a beam of ions (electrically charged atoms) is shot through a glowing gas. The beam is found to curve as it travels through the gas, and it is found that this deflection depends on exactly where the beam is shot from. The graph shows the relationship between two variables: the independent variable is the original position of the beam, and the dependent variable is the amount of deflection that results. Now because of certain uncontrolled variables present in this experiment, the beam deflection has an uncertainty of ±0.2 mm, which is fairly large when compared with a measurement result on the order of a few millimetres. Since the uncertainties here are so large, they are plotted on the graph right along with the data points. The vertical bars extend 0.2 mm above and 0.2 mm below the data points, indicating the uncertainty of ±0.2 mm in the measurements of the dependent variable.

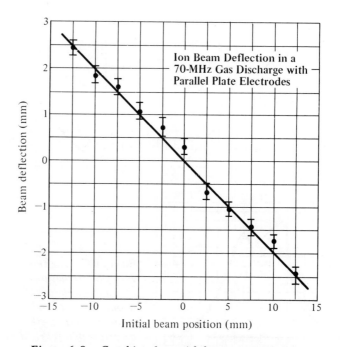

Figure 6-8. *Graphing data with large uncertainties.*

Interpolation with Uncertainty Bars

One important feature of figure 6-8 is the interpolation. This interpolation is a straight line drawn to intersect some part of the uncertainty range of each data point. If it were not possible to do this with a straight line, then a more complicated curve would have had to be drawn. The point is that with such large uncertainties, it makes no sense to draw a line that goes precisely through the center of each data point. As long as the line goes somewhere through the uncertainty range of each point, it is totally consistent with the data. In fact, even if the line makes a near miss or two, it is still acceptable.

"But," you say, "there may be more than one line that can be drawn this way. How do we know we have the right one?" The answer is that we do not know. If the data has uncertainties, then the line drawn to interpolate the data also has uncertainties. A graph, after all, can be no more accurate than the data it represents.

Drawing the best line consistent with the data is a matter of skill combined with judgment. A few rules of thumb should be kept in mind, however:

Rules for Graphical Interpolation

1. When uncertainties are large enough to show up on the scale of the graph, it is not necessary to force the line to go through the actual data points, even the end points.
2. Never draw a curved line when a straight line can be drawn consistent with the same data.
3. If a curved line must be drawn, it should be the simplest possible curve. That is, it should have no cusps or abrupt jumps, and it should be as smooth as possible.
4. If you have many points and if the data has large uncertainties, the number of points the line misses on one side should roughly counterbalance the number of points missed on the other side.

The Squint Test

One way of judging a curve's smoothness is to give it the so-called squint test. Holding your eye close to the plane of the graph, keep the paper as flat as possible, and sight along the curve as shown in figure 6-9. From this vantage point, any cusps or kinks in the curve become immediately apparent.

Why so much fuss about the smoothness of the curves we draw? Simply

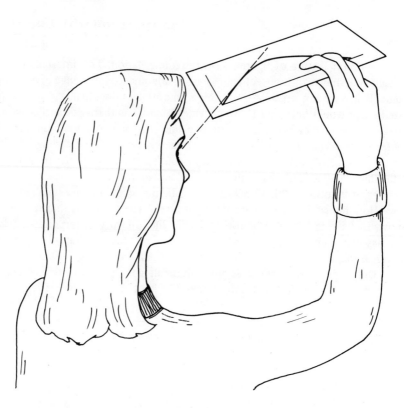

Figure 6-9. *The "squint test" for checking the smoothness of a curve. By sighting close to the plane of the paper, any cusps or kinks become immediately obvious.*

this: in nature, the relationship between two variables is almost always smooth. The only major exceptions to this are situations involving phase changes (freezing or boiling) and events that take place inside the atom. So in the vast majority of cases, any jumps or kinks in curves are guaranteed to be wrong. The smoother the curve, the more likely it is that it truly represents the relationship between the quantities being graphed.

Combining Graphs

Often, for purposes of comparison, we will want to draw more than one curve on the same graph. An example of this is shown in figure 6-10, where the boiling point curves of four different liquids are displayed at once. In actuality, then, figure 6-10 is four separate graphs.

The advantage of putting all these curves on one grid, however, amounts to

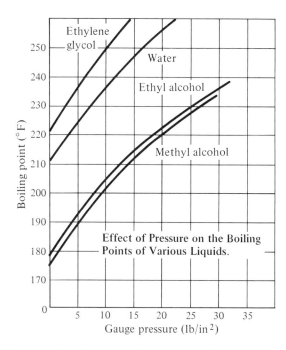

Figure 6-10. *A graph with four dependent variables.*

more than a matter of saving space. In one glance, we can compare the properties of these four liquids. We can see, for example, that the boiling point of ethyl alcohol is always a few degrees higher than the boiling point of methyl alcohol at the same pressure. We can see that water must be pressurized to about 4 lb/in² to have the same boiling point as ethylene glycol at atmospheric pressure. Likewise, we can see that methyl alcohol at a pressure of about 23 lb/in² has the same boiling point as water at 5 lb/in² pressure. And so on. It would be very inconvenient to try to make comparisons like these from four separate graphs.

In figure 6-10, the four curves represent the behavior of four different substances. In this case, we actually have four different dependent variables with only one independent variable. There is really no limit to the number of dependent variables that can be shown on the same graph, other than the fact that eventually the graph will begin to get too crowded.

Occasionally, we may want to draw a graph representing the relationship between one dependent variable and two independent variables. Figure 6-11 is an example. Here the dependent variable is the energy density of the thermal radiation from a perfect radiator (a so-called blackbody). This energy density is found to be different at different wavelengths; hence the wavelength is an independent variable. It is also found, however, that the whole relationship between these two variables is different at different temperatures. Temperature, then, is a second independent variable. One way of showing the entire relationship is to draw several curves, each one corresponding to a single temperature. We can see, then, how the energy density depends on wavelength *and* how it

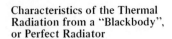

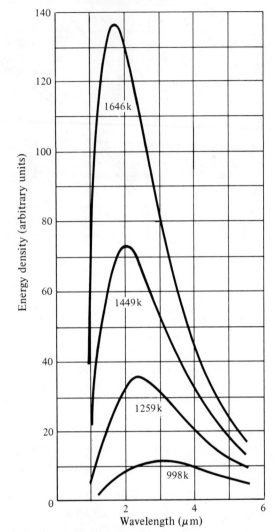

Figure 6-11. *A graph with one dependent variable and two independent variables (wavelength and temperature).*

depends on temperature. Let's take a closer look at the steps involved in drawing such a graph.

Constructing a Graph with Two Independent Variables

▶ **Example 6-5: Magnetic field of a current loop**
Data is taken on the strength of a certain magnetic field at different points

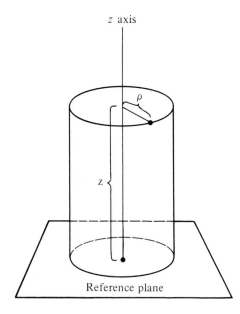

Figure 6-12. *Cylindrical coordinates for locating points in space. Points lying on the upper circle are located by specifying z, the vertical height from the reference plane, and ρ, the radial distance from the z axis.*

Table 6-2. *Value of the Magnetic Field at Different Points in Space (in Tesla)*

		ρ (cm)			
		0.25	1.75	2.75	3.75
z (cm)	0.00	0.00	0.00	0.00	0.00
	0.50	0.133	0.257	0.543	1.770
	1.00	0.253	0.474	0.926	2.356
	1.50	0.353	0.625	1.106	2.207
	2.00	0.426	0.709	1.135	1.874
	2.50	0.474	0.739	1.081	1.556
	3.00	0.499	0.732	0.992	1.295
	3.50	0.505	0.701	0.894	1.089
	4.00	0.498	0.659	0.799	0.926
	4.50	0.481	0.611	0.713	0.796

in space. The points are located by two space coordinates: z, the vertical height above a certain reference plane, and ρ, the radial distance from a certain line (the z axis) drawn perpendicular to this plane (figure 6-12). Table 6-2 shows the value of the magnetic field, in units of tesla (T), at these different points.*

---

*More precisely, the data represents the ρ component of the magnetic flux density resulting from a single current-carrying loop of wire in the reference plane. The loop is 10 cm in diameter. Although this data was actually generated by computer, similar kinds of data can result from direct measurement.

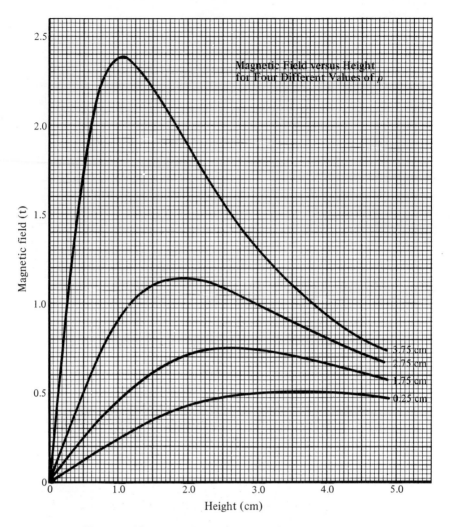

Figure 6-13. *A graph with two independent variables.*

The problem is to represent this data on a graph. Since there are two independent variables (the two space coordinates), we have no choice but to draw more than one curve.

Now, we have two ways of doing so. One is to graph the magnetic field versus ρ, then to repeat this for the 10 different values of z. This would give us a graph with 10 curves on it. Not only would such a graph be confusing to look at, but since each curve would have only four points on it, none of the interpolations would be very accurate.

A better approach is to plot the magnetic field versus z, then to repeat this for the four different values of ρ. This gives four curves, with 10 points on each curve. To keep the curves from getting confused with one another, it is still best to plot them one at a time. The result is shown in figure 6-13. Notice that each curve is labeled with its particular value of ρ. ◀

Graphs without Grids

It is fairly common practice, in books and technical reports, to reproduce graphs without the background grid pattern. Figure 6-14 is an example. Unfortunately, it is easy to get the impression that such a graph lacks accuracy. In fact, such is not the case. Graphs like figure 6-14 are traced from graphs with full grid patterns, and should be every bit as accurate as the graphs they are traced from. The grid is omitted to allow for a more pleasing, uncluttered visual impression. Often it is left off when a graph is used to display a typical data trend rather than a particular set of data values. In any case, if the graph is expected to see a fair amount of use, there is no good justification for omitting the grid.

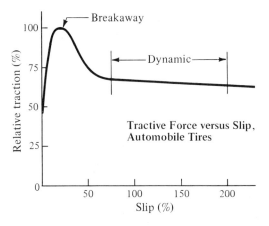

Figure 6-14. *A graph with no grid. Such graphs are usually used to display general behavior rather than a specific set of data values.*

Summary

A graph is used to compress data into a small space, while making it easier to refer to than in a table. While they offer advantages in comparing data values and demonstrating trends in the data, graphs suffer the disadvantage of sacrificing precision.

Graphs may be drawn from either statistical or functional tables, but functional graphs are the more common. The *dependent variable* should be plotted on the vertical axis, and the *independent variable* should be plotted on the horizontal axis. Because a larger graph gives better precision, the scale should be chosen so the graph comes reasonably close to filling the paper. If the data has large uncertainties, these should be indicated by bars drawn through the data points. *Linear interpolation* is done with a straightedge, while *nonlinear inter-*

polation should be done with a french curve. The interpolated graph should have no cusps or abrupt jumps.

By drawing more than one graph on the same grid, the physical behaviors of different substances can be compared easily. A dependent variable that depends on two independent variables may also be represented by a series of curves on the same grid.

Although drawing the graph is usually the final step in the measurement process, sometimes a measurement is based on previously graphed data. This is the case with many indirect measurements. Such measurements are given special consideration in the next chapter.

REVIEW QUESTIONS

1. What is a graph?
2. What are some of the advantages graphs have over tables as a means of representing data?
3. What is the principal disadvantage of graphs?
4. What kinds of graphs can be drawn from statistical tables?
5. Outline the procedure for drawing a bar graph.
6. Outline the procedure for drawing a circle graph.
7. What type of graph is based on a functional table?
8. What is a dependent variable?
9. What is an independent variable?
10. Outline the procedure for drawing a graph from a functional table.
11. What is the process of connecting data points on a graph called?
12. What is a french curve used for?
13. How are data uncertainties represented on a graph?
14. Why should cusps and abrupt jumps be avoided in graphical interpolations?
15. What is the "squint test"?
16. There are two reasons we might want to draw more than one graph on the same grid. What are they?
17. Why is the background grid sometimes omitted from a graph?

EXERCISES

1. An acceleration test gives data on the speed of a car at different instants in time. The highest speed in the data is 176 km/h, attained 33 s after the car starts. The entire set of data is to be graphed on a piece of graph paper having 100 vertical and 80 horizontal divisions.
   a. What is the dependent variable?
   b. What vertical scale should be used? (*Answer:* 1 division = 2 km/h)
   c. What horizontal scale should be used?

2. An engine's power output depends on its frequency of rotation. Data is obtained on the variation in power at frequencies between 0 and 6 000 r/min. The highest power measurement is 146 hp. The data are to be graphed on a piece of graph paper having 250 vertical and 180 horizontal divisions.
   a. What is the dependent variable?
   b. Should the paper be used horizontally or vertically?
   c. What vertical scale should be used?
   d. What horizontal scale should be used?

3. The following list gives the melting points of some low-melting alloys. Represent this data on a bar graph.

Wood's metal	70°C
Newton's metal	97
Onion's metal	100
Malotte's metal	123
soft solder	192
white metal	199

4. There are five naturally occurring isotopes of zinc. The natural abundance of each of these isotopes is listed below, expressed as a percent of all zinc found in the earth. Draw a circle graph to represent the data.

Mass Number	Natural Abundance (%)
64	48.89
66	27.81
67	4.11
68	18.57
70	0.62

5. The ultimate strength of Portland-cement concrete depends on its age. For concrete that has been subject to standard moist curing and is placed under compression, the ultimate strength is given in the following table. Graph the data.

Constructing Graphs

Ultimate Strength (lb/in^2)	Age (mo)
0	0
2 200	0.3
3 300	1.0
3 800	2.0
4 200	3.0
4 800	10.0

6. Thermal conductivity is the rate at which heat energy flows through a unit cross-sectional area of a substance of unit thickness, when the temperature difference across the sample is one degree. The thermal conductivity of a given substance will depend somewhat on the temperature at which it is measured. The following table gives the thermal conductivity of dry air at different temperatures. Graph the data.

Thermal Conductivity [(cal/cm^2·s)/($^\circ$C/cm)]	Temperature ($^\circ$C)
2.1×10^{-5}	-180
3.9×10^{-5}	-100
4.9×10^{-5}	-50
5.76×10^{-5}	0
6.1×10^{-5}	20
7.4×10^{-5}	100
8.8×10^{-5}	200

7. The following table gives the density of mercury as a function of temperature. Graph the data. (*Hint:* The density axis need not start at zero.)

Temperature ($^\circ$C)	Density (g/cm^3)
-20	13.644 6
0	13.595 1
20	13.545 8
40	13.496 9
60	13.448 2
80	13.399 8
100	13.351 5

8. The percent of incident light transmitted by an optical filter depends on the light's wavelength. The following data are for three different types of Wratten filter manufactured by the Eastman Kodak Company. Graph all these data on one graph. (*Hint:* It may be a good idea to plot each set of data with a different color of pen.)

Wavelength (μm)	Percent Transmittance		
	No. 52	No. 61	No. 66
400	2.18	—	12.3
420	0.80	—	15.0

440	0.41	–	23.2
460	1.45	–	42.2
480	4.90	0.33	68.4
500	13.3	16.6	82.7
520	23.7	40.0	84.0
540	32.1	34.5	79.1
560	31.0	17.3	67.1
580	19.1	4.40	47.2
600	7.78	0.38	24.4
620	2.34	–	13.7
640	0.80	–	3.00
660	0.36	–	1.91
680	0.23	–	19.9
700	0.17	–	63.1

9. The force needed to feed a twist drill depends on the material being drilled, the desired feed rate, and the size of the drill. Approximate values for newly sharpened drills drilling cast iron are given in the following table. Graph the data.

Feed per Revolution (in.)	Drill Diameter (in.)			
	1/4	1/2	3/4	1
0.004	135	200	275	350
0.006	200	300	410	520
0.008	265	400	545	690
0.010	330	500	680	850
0.012	395	595	805	1 005
0.014	460	690	925	1 155
0.016	520	780	1 040	1 300

10. If the volume of air fed into a coal-fired boiler exceeds the amount needed for complete combustion of the coal, the flue gas will contain some unused oxygen. The following table lists the percent of oxygen by volume in the flue gas for different amounts of excess air. Graph the data.

Oxygen in Flue Gas (% by volume)	Excess Air for Combustion (% by volume)
0	0
1.9 ± 0.1	10.0
3.5 ± 0.2	20.0
4.9 ± 0.25	30.0
7.1 ± 0.3	50.0
8.8 ± 0.4	70.0
10.2 ± 0.5	90.0
11.2 ± 0.5	110.0

CHAPTER 7

The Propagation of Uncertainty

There is a famous story that goes something like this: For the want of a nail, the shoe was lost. For the want of a shoe, the horse was lost. For the want of a horse, the battle was lost. And so on until the war, and finally the kingdom was lost. All because a little nail was missing from a horseshoe.

Propagation of Errors in Multistep Procedures

This old story illustrates one basic fact of nature: that errors tend to *propagate*, or repeat themselves. If an error is introduced at one step in a process, it will appear in each subsequent step, and very often, its effect will increase from one step to the next. As a result, a relatively small error made at the beginning of a multistep process can become very sizable.

A familiar example is shown in figure 7-1. Here we see a standard 4-ft by 8-ft pool table. The object ball is to be sunk in the side pocket after banking it from three rails. The dotted line shows the ball's theoretical path, assuming a perfect shot with no "English" (angular momentum). The actual shot is shown with the solid line. The initial error here is about 1 angular degree. As a result, the first bank occurs about 1 in. (2.5 cm) from where it should. By the time the ball hits the second rail, this error has increased to over 3.7 in. (9.4 cm). At the third rail, it has increased to around 3.9 in. (9.9 cm), and the ball finally misses the side pocket by a whopping 9 in., or nearly 23 cm! On this basis, it is a terrible shot. The error at the first rail, though, was only one inch. This error grew considerably as the ball rolled along.

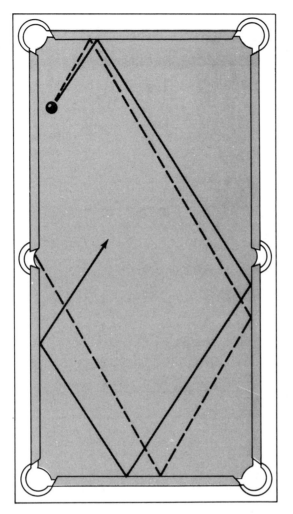

Figure 7-1. *Propagation of error on a pool table. An error of 1 in. at the first rail grows to 9 in. at the fourth rail.*

Calculating with Uncertainties

Our main concern here, however, is measurement uncertainty rather than shooting pool. Many measurements involve some calculation, and nearly all measurements are made with the idea of using the results in calculations of one sort or another. What we need to examine, then, is how to do calculations with quantities that have uncertainties. And we will have to pay particular attention to how these uncertainties propagate during the calculations.

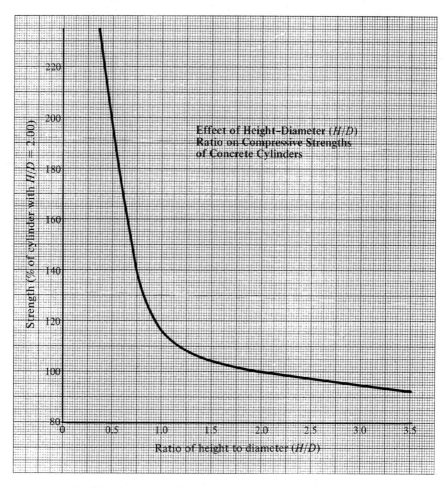

Figure 7-2. *Factors such as age and composition being equal, the strength of a concrete support column depends on its geometry. The relative strength of such a cylindrical column may be determined by measuring the ratio of height to diameter and consulting a graph such as this one. Notice that the strength is given relative to a "standard" column whose height is twice its diameter.*

The simplest type of calculation is reading a graph. Figure 7-2 shows how the strength of a concrete column depends on the ratio of its height to its diameter (H/D). Tall, narrow columns have large values of H/D, and the graph shows that they are not very strong. Short, stubby columns tend to be much stronger. The strength here is expressed as a percent of the strength of a "standard" column having an H/D ratio of 2.00. We assume that the columns being compared are of equal composition and age.

Indirect Measurement

Suppose we need to know the relative strengths of several concrete columns that are already part of a structure. We cannot very well take the structure apart to run tests. We can, however, measure the height and diameter of each column, calculate the *H/D* ratios, then look at the graph to get the relative strengths. In effect, we use the graph to *calculate* the strengths from the *H/D* ratios. We still make the strength measurement, but we do so indirectly.

Now comes the problem. There is no way we can measure the *H/D* ratios with perfect accuracy. Maybe the columns are slightly tapered. Maybe they have bases or capitals—something the graph does not tell us how to treat. The cross section is probably not a perfect circle, and the actual cross-sectional shape may vary a bit from one end to the other. Since *H/D* has an uncertainty, the strength will have an uncertainty also. Example 7-1 shows how to find it.

Determining Uncertainty in an Indirect Measurement

▶ **Example 7-1: The strength of concrete columns**

A certain structure has three different sizes of concrete columns. The ratios of their heights to diameters are measured, with the results shown below. Find the relative strengths of the columns, and the uncertainties in these figures.

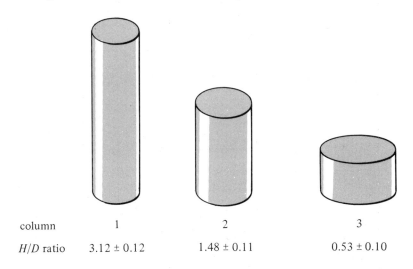

column	1	2	3
H/D ratio	3.12 ± 0.12	1.48 ± 0.11	0.53 ± 0.10

We proceed as follows: The first column has an *H/D* ratio that falls between 3.00 and 3.24. Consulting the graph, we see that 3.00 corresponds to a strength of 93%, while 3.24 corresponds to nearly 95%. The range of possible strengths, then, is 93% to 95% of the strength of the

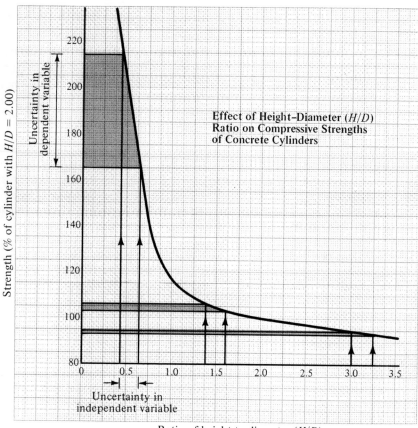

Figure 7-3. *The uncertainty in the dependent variable depends on the uncertainty in the independent variable, and may be read directly from a graph if one is available.*

"standard" column, or 94% ± 1%. Figure 7-3 shows this H/D range and the corresponding range in strength. The graph itself is the same as the graph in figure 7-2.

The second column has an H/D range between 1.37 and 1.59. The corresponding range in strength is seen to be 103% to 106%, or 104.5% ± 1.5%.

The third column, the stubbiest, has an H/D ratio falling between 0.43 and 0.63. From figures 7-2 and 7-3, we see that this ratio corresponds to a strength of between 165% and 214%, or 189.5% ± 24.5%. We are justified in rounding this off to 190% ± 25%.

Notice that the graph has given us not just the relative strengths of these three columns, but also the uncertainties in these figures. Notice also that the uncertainty in strength is much greater for the third column than for the first two, even though the uncertainties in H/D were nearly the same in all three cases.

100 / Chapter 7

Summarizing the results, we have:

Column	H/D	Strength
1	3.12 ± 0.12	94% ± 1%
2	1.48 ± 0.11	104.5% ± 1.5%
3	0.53 ± 0.10	190% ± 25%

Examine these results carefully and note how they are a natural consequence of the original graph in figure 7-2. ◀

In the preceding example, the uncertainty in the dependent variable was a direct result of the uncertainty in the independent variable. The graph itself did not add to this uncertainty because it had been drawn very accurately and very precisely.

Graphs That Increase the Uncertainty

Occasionally, as we saw in chapter 6 (figure 6-8), a graph may not be as accurate as the grid it is drawn on. Often, such graphs are drawn as a heavy band rather than as a single line. Figure 7-4 is an example. Here we see the yield strength of steel as a function of its hardness. The hardness is determined here by the Brinell test, which is based on pressing a tungsten carbide ball into the sample's surface, then measuring the size of the impression. The device is shown in figure 7-5. The yield strength is the stress under which the steel begins to flow plastically, which is also the minimum stress that will produce a permanent deformation. Measuring the yield strength directly involves destroying the sample, while measuring the Brinell hardness does not. In fact, the hardness can be measured (albeit with some difficulty) on a sample in an existing structure. Because it is difficult to detect the yield point exactly, there is a fair amount of uncertainty in this quantity. Thus the graph itself is drawn with a band of uncertainty.

Determining Uncertainty from an Uncertain Graph

▶ **Example 7-2: Yield strength from a hardness measurement**
The yield strength needs to be found for a certain sample of steel, but the sample itself is not to be destroyed. The measurement is therefore made indirectly, by measuring the Brinell hardness, then by using the graph in figure 7-4 to obtain the yield strength. Suppose that the Brinell hardness comes out as 440 ± 20. What is the yield strength and its uncertainty for this sample?
The procedure is basically the same as that in example 7-1. The Brinell

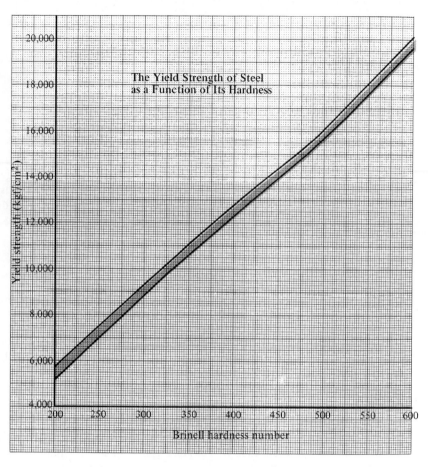

Figure 7-4. *A graph with a built-in uncertainty. The yield strength of a steel sample may be found indirectly by measuring its hardness, then consulting this graph. The graph itself adds to the uncertainty in the hardness measurement.*

hardness has an uncertainty range of 420 to 460. The corresponding range in yield strength is read from the graph. In doing so, we must take into account the uncertainty in the graph itself. We figure the uncertainty by taking the upper limit in yield strength from the upper limit hardness and the *top* of the curve; we find the lower limit in yield strength from the lower limit hardness and the *bottom* of the curve (figure 7-6).

The result is that the yield strength is somewhere between about 13 000 kilogram-force per square centimetre (kgf/cm^2) and 14 700 kgf/cm^2. In the usual way, we write this as

$$\text{yield strength} = 13\,850 \pm 850 \text{ kgf/cm}^2$$

Can a measurement with this much uncertainty be good for anything? It certainly can. For example, if the sample is to be used in a structure where its strength must be greater than 12 000 kgf/cm^2, the measurement

Figure 7-5. *A Brinell hardness tester. (Courtesy of Detroit Testing Machine Co.)*

indicates that this sample passes the test. If, on the other hand, the sample is to be used in a forming application, then its yield strength might have to be less than, say, 9 000 kgf/cm². If that is the case, then this sample must be rejected (or at least annealed to make it softer). Obviously, in either case it is important to know the uncertainty in the measurement. ◄

Propagation of Uncertainty

Let's summarize the main point of these last two examples. If we know how to calculate one quantity from another, then we have a means of finding the one

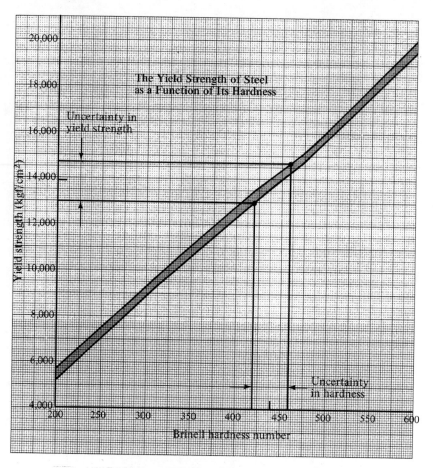

Figure 7-6. *Determining uncertainty from a graph that itself has uncertainty. A Brinell hardness of 440 ± 20 is found to correspond to a yield strength of 13 850 ± 850 kgf/cm².*

quantity by measuring the other one. Equally important, we can use the same relationship to find the uncertainty in the calculated quantity from the uncertainty in the measured quantity. The original measurement uncertainty *propagates*, or reproduces itself, in the second quantity. And just as children may grow to different sizes than their parents, so this propagated uncertainty may be a different size than the uncertainty in the original measurement.

DEFINITION | *Propagation of uncertainty* is the process by which the uncertainty in a measured quantity is carried along by a calculation made with the quantity.

Let's now look at some examples of the propagation of uncertainty in arithmetic calculations.

Uncertainty in a Function of a Single Measured Variable

▶ **Example 7-3: Vertical distance using falling objects**
The laws of falling bodies can be used to predict the distance that a freely falling object will travel in a given time. If an object is dropped from rest and allowed to fall for a time t, then the distance it falls, y, is given by

$$y = \tfrac{1}{2}gt^2$$

where g is the acceleration due to gravity. Sea level values for this quantity were listed in table 5-3.

Early cave explorers often used this relationship to estimate the depth of pits and shafts they encountered in caves. If the pit was too deep for them to see the bottom (their portable lighting was not very bright), they would toss in a stone and measure the time lapse before the thud or splash from the bottom. A quick calculation then gave the depth. Such a method was a great deal easier than lowering a rope each time.

Suppose that the time of fall is measured as 2.8 ± 0.2 s. How deep is the pit?

Let's say that the latitude is 35°N, and that we are not far from sea level. The acceleration due to gravity, from table 5-3, is

$$g = 32.143\ 5\ \text{ft/s}^2$$

The depth of the pit, then, is

$$y = \tfrac{1}{2}(32.143\ 5\ \text{ft/s}^2)(2.8\ \text{s})^2$$

$$y = 126.00\ \text{ft}$$

To get the uncertainty, we reason as follows: The stone could have been falling for as long as 2.8 + 0.2, or 3.0 s. If so, it had to fall a distance of

$$y = \tfrac{1}{2}(32.143\ 5\ \text{ft/s}^2)(3.0\ \text{s})^2$$

$$y = 145\ \text{ft}$$

On the other hand, it might have been falling only 2.8 - 0.2, or 2.6 s. If that had been the case, it would have fallen a distance of

$$y = \tfrac{1}{2}(32.143\ 5\ \text{ft/s}^2)(2.6\ \text{s})^2$$

$$y = 109\ \text{ft}$$

These calculations give three different values to work with:

upper limit	145 ft
best value	126 ft
lower limit	109 ft

Since the 126 ft followed from the original 2.8-s measurement, it is the "best value" for the result. The other two values define the uncertainty limits. Notice that the upper limit is 19 ft higher than the best value, and the lower limit is 17 ft lower than the best value. The average error limit is then 18 ft, and we can quote the result as

$$y = 126 \pm 18\ \text{ft}$$

The accuracy of this procedure is limited by the difficulty of making precision time measurements in the field. A small uncertainty in time has a very large effect on the accuracy of the result. We might also point out that if the stone falls a very great distance, air resistance will begin to have an effect, and the equation we used here will no longer be accurate. ◄

The last example involved only one independent variable. In most situations, however, there will be two or more. Examples 7-4 and 7-5 show how to handle such cases.

Uncertainty in a Product of Two Measured Variables

► **Example 7-4: Fluid force from pressure and area measurement**
The pressure exerted by a fluid is defined as the force per cross-sectional area. In symbols,

$$P = \frac{F}{A}$$

where F is the force, A is the area the force acts on, and P is the pressure.

In hydraulic applications, we often find that we can easily measure the fluid pressure and the area of a pipe or a piston, but it is the force that we need to know. If so, we can solve for F to get

$$F = PA$$

To get the force, then, we need to make *two* measurements, then multiply the results together.

Suppose that the pressure and area measurements give the following results. What is the force and its uncertainty?

$$P = 8\ 563 \pm 74 \text{ pascals (Pa)}$$

$$A = 0.076\ 2 \pm 0.001\ 2 \text{ m}^2$$

The best value is easy to get, since it is the product of the two best values. This gives

$$F = (8\ 563 \text{ Pa})(0.076\ 2 \text{ m}^2)$$

$$F = 652.5 \text{ newtons (N)}$$

This result is in units of *newtons*, the SI unit for force, because the pascal is the SI unit for pressure, and the square metre is the SI unit for area. Chapter 13 will have more to say about relationships between units; for now, suffice it to say that $1 \text{ N} = 0.224\ 8 \text{ lb}$.

Now, the pressure might actually have been higher than 8 563 Pa; in fact, it could have been as high as 8 637 Pa (8 563 + 74). At the same time, the area could have been as high as 0.077 4 m². If so, then the force works out as

$$F = (8\ 637\ \text{Pa})(0.077\ 4\ \text{m}^2)$$
$$F = 668.5\ \text{N}$$

This is the upper limit on the force.

To get the lower limit, we use the lower limit on the pressure and the lower limit on the area:

$$F = (848\ 9\ \text{Pa})(0.075\ 0\ \text{m}^2)$$
$$F = 636.7\ \text{N}$$

So far, then, we have the following three values:

upper limit	668.5 N
best value	652.5 N
lower limit	636.7 N

The upper limit is 16.0 N higher than the best value, while the lower limit is 15.8 N lower than the best value. Since it makes little sense to quote an uncertainty to three-digit precision, we can round these results off to a final answer of

$$F = 653 \pm 16\ \text{N}$$

We can summarize the entire procedure as follows: ◀

Procedure for Finding Uncertainty in Calculated Quantities

1. Look up the correct formula for calculating the unknown quantity from one or more independent variables.
2. Determine the value of each independent variable and its uncertainty. This usually means that you must make a measurement, or use the results of someone else's measurement.
3. Calculate the best value of the dependent variable, using the formula and the best values of the independent variables.
4. Calculate the upper limit of the dependent variable, using the same formula and those limits of the independent variables that contribute to a high result.
5. Repeat step 4 for the lower limit of the dependent variable.
6. The total uncertainty range is the different between the answers found in steps 4 and 5. Half this difference is the uncertainty in either direction from the best value.

We have to be extra careful in steps 4 and 5, since the upper limit of the dependent variable may not always result from the upper limits of the independent variables. Example 7-5 shows such a case.

Uncertainty in a Ratio of Two Measured Variables

▶ **Example 7-5: Speed from distance and time measurement**
The measurement of speed (or velocity) usually involves making separate measurements of distance and time. The speed is then calculated by dividing one by the other:

$$v = \frac{x}{t}$$

where x is the distance, t is the time, and v is the speed or velocity.

Suppose that the underwater speed of sound is being measured by having the sound travel a certain distance. The time and distance measurements are

$$x = 324 \pm 1 \text{ m}$$
$$t = 0.227 \pm 0.002 \text{ s}$$

What is the speed and its uncertainty?

We calculate the best value of the speed from the two best values. This gives

$$v = \frac{324 \text{ m}}{0.227 \text{ s}}$$

$$v = 1\ 427 \text{ m/s}$$

Now possibly the distance was really a little longer than we thought; if so, then the speed will come out higher. The time might also have been slightly in error. Notice, however, that if the time is larger, then the calculated speed is *smaller*. So if we are looking for the *upper limit* of the speed, then we must use the *lower limit* of the time.

If this does not seem to make sense, think about this: You take a certain trip and it takes you 5 hours. A friend takes the same trip, but it takes her only 4 hours. Who had the greater average speed? Obviously your friend did, for she had a shorter travel time.

Mathematically, it amounts to this: If the denominator of a fraction gets smaller, the fraction itself gets larger. The fraction 2/5 gets larger if the 5 in the denominator is replaced by a 3, and it gets larger still if the 3 is replaced by a 2.

So the upper limit of the speed is found by using the largest distance and the shortest time. This gives

$$v = \frac{325 \text{ m}}{0.225 \text{ s}}$$

$$v = 1\ 444 \text{ m/s}$$

In a similar way, we find the lower limit by using the shortest distance and the largest time.

$$v = \frac{323 \text{ m}}{0.229 \text{ s}}$$

$$v = 1\,410 \text{ m}$$

The difference between the high and the low is 34 m/s. Half of this is 17 m/s, which is to say that the uncertainty range is an average ±17 m/s around the best value. The final result, then, is

$$v = 1\,427 \pm 17 \text{ m/s}$$

Suppose that this uncertainty is too large to be acceptable. We want to repeat the original measurements and improve on them. Which measurement should we concentrate on, the distance or the time? We can answer this kind of question by repeating the calculations—first with zero uncertainty in the distance, then with zero uncertainty in the time. The details will be left to you. The result is that even with a perfect distance measurement, the uncertainty would still be ±10.5 m/s. So the very small uncertainty of ±0.002 s in time eventually makes a very large contribution to the final uncertainty. If we can do nothing to improve this time measurement, it makes little sense to spend much effort improving the distance measurement. ◄

Direct Calculation of Propagated Uncertainty

We can always determine the uncertainty in a calculated quantity by the procedure we have just developed. There is also another way, which is sometimes a bit quicker. It has been a secret until now because it is really important that you be able to do things the first way. The second method depends on the use of table 7-1, which may not always be handy.

Suppose that we have a dependent variable z that depends on the independent variables x and y according to

$$z = xy$$

We make measurements on x and y and find that

$$x = 98.6 \pm 2.1$$

$$y = 0.217 \pm 0.003$$

We want to find the uncertainty in z. In table 7-1, this uncertainty is symbolized by Δz. Its value is given by the formula

$$\Delta z = x\,\Delta y + y\,\Delta x$$

where Δx and Δy are the uncertainties in x and y. Numerically, this gives us

$$\Delta z = (98.6)(0.003) + (0.217)(2.1)$$

$$= 0.296 + 0.456$$

$$\Delta z = 0.752$$

Table 7-1. *Formulas for the Propagation of Uncertainty*

Formula	Uncertainty
$z = kx$	$\Delta z = k\,\Delta x$
$z = x + y$	$\Delta z = \Delta x + \Delta y$
$z = x - y$	$\Delta z = \Delta x + \Delta y$
$z = xy$	$\Delta z = x\,\Delta y + y\,\Delta x$
$z = \dfrac{x}{y}$	$\Delta z = \dfrac{x\,\Delta y + y\,\Delta x}{y^2}$
$z = x^2$	$\Delta z = 2x\,\Delta x$
$z = x^3$	$\Delta z = 3x^2\,\Delta x$
$z = \sin kx$	$\Delta z = k\,\Delta x \cos kx$
$z = \cos kx$	$\Delta z = k\,\Delta x \sin kx$
$z = \sqrt{kx}$	$\Delta z = \dfrac{k\,\Delta x}{2\sqrt{kx}}$

Note: If the quantities are related by a formula of the form given in the left column, then the uncertainty can be calculated from the corresponding formula on the right. The symbols are defined as follows:

- z = dependent variable
- x, y = independent variables that are measured directly
- k = any constant (e.g., 2, π) not requiring measurement
- Δz = uncertainty in dependent variable
- $\Delta x, \Delta y$ = uncertainty in independent variables

which we may round off to 0.75. This is exactly what we would get according to the method in example 7-4.

Let's look at an example where the symbols are not x, y, and z.

Using Table 7-1 to Calculate Uncertainty

▶ **Example 7-6: The speed of sound**

In example 7-5, we saw that the speed of sound can be found from

$$v = \frac{x}{t}$$

and we had the data

$$x = 324 \pm 1 \text{ m}$$
$$t = 0.227 \pm 0.002 \text{ s}$$

Let's find the uncertainty in v by using table 7-1.

We begin by looking through the table for a case where one variable is divided by another. The one we find is

$$z = \frac{x}{y}, \quad \Delta z = \frac{x \Delta y + y \Delta x}{y^2}$$

Now this is basically what we have, except that we have v instead of z, and t instead of y. If we make these substitutions, we get

$$v = \frac{x}{t}, \quad \Delta v = \frac{x \Delta t + t \Delta x}{t^2}$$

Then putting in the numbers, we find

$$\Delta v = \frac{(324)(0.002) + (0.227)(1)}{(0.227)^2}$$

$$= \frac{0.648 + 0.227}{0.0515}$$

$$\Delta v = 16.98$$

This result rounds off to the 17 = m/s uncertainty we calculated previously.

Notice that it is now very easy to see what happens if we reduce the uncertainty in one of the variables. Suppose that x is measured very accurately, so that Δx is practically zero. Then

$$\Delta v = \frac{(324)(0.002) + (0.227)(0)}{(0.227)^2}$$

$$= \frac{0.648}{0.0515}$$

$$\Delta v = 12.58$$

This means that the original uncertainty in time contributes over 12 m/s of the total uncertainty. This formula method, then, is a fairly straightforward way to analyze where the major errors are coming from. ◀

Using Table 7-1

▶ **Example 7-7: Speedometer calibration using a tachometer**

A car's speedometer can be calibrated against its tachometer, if it has one. This is the basic procedure.

The car's forward speed is determined by the size of its tires and the frequency of rotation of its drive wheels. If f is the final drive frequency (in revolutions per minute), and d is the diameter of the drive tires (in feet), then the forward speed of the vehicle (in miles per hour) is

$$v = \frac{\pi}{88} fd$$

The number 88 in this formula is exact; so, of course, is π. The formula assumes no slippage between the tires and the road.

Now the final drive frequency, f, can be found by dividing the tachometer reading by the overall drive ratio for the particular car. Suppose that this result comes out to be

$$f = 933 \pm 29 \text{ r/min}$$

The uncertainty here results mainly from the tachometer scale's 3% accuracy.

The tire diameter may come out as

$$d = 2.068 \pm 0.042 \text{ ft}$$

This uncertainty comes mainly from the difficulty in estimating how much the tire expands when it is rolling.

Under these conditions, the speedometer reading should be

$$v = \frac{\pi}{88} fd$$

$$= \frac{\pi}{88} (933)(2.068) \text{ mi/h}$$

$$v = 68.88 \text{ mi/h}$$

We now need to find the uncertainty in this figure. The closest formula in the table is

$$z = xy$$

If we make the obvious replacements, we might be left wondering what to do with the $\pi/88$. Surely it belongs somewhere in the uncertainty formula.

And it does. The first formula in table 7-1 is

$$z = kx, \qquad \Delta z = k \, \Delta x$$

which tells us that if there is a constant factor in a formula, this same constant factor appears in the uncertainty. It is a perfectly general mathematical statement, and it applies in the present case. So we have

$$z = kxy, \qquad \Delta z = k(x \, \Delta y + y \, \Delta x)$$

where we will replace z by v, x by f, y by d, and k by $\pi/88$. This gives

$$v = \frac{\pi}{88} fd, \qquad \Delta v = \frac{\pi}{88} (f \, \Delta d + d \, \Delta f)$$

Now we can do the numerical calculation.

$$\Delta v = \frac{\pi}{88} [(933)(0.042) + (2.068)(29)] \text{ mi/h}$$

$$= \frac{\pi}{88} (99.16) \text{ mi/h}$$

$$\Delta v = 3.54 \text{ mi/h}$$

The final result, then, is that the original tachometer reading corresponds to a forward speed of

$$v = 68.9 \pm 3.5 \text{ mi/h}$$

This uncertainty may seem disappointingly high, but actually it is not. It represents an uncertainty of about 5%, and the dial-type tachometer and speedometer themselves can be read only to about 3%. It is also consistent with the 5 mi/h legal tolerance.

How can this calibration be improved? Someone is bound to suggest running a measured mile. This is fine, but such a measurement would give only the *average* speed for the mile, not the actual speed corresponding to a given engine frequency. The driver is bound to vary his or her speed slightly over such a course, and gusts of wind can make the variation greater. Moreover, if the car is not driven in a perfectly straight path (if it swerves from side to side during the run), then it actually travels slightly farther than a mile. So the measured mile is no great improvement over the technique described here, and it has the disadvantage that a measured mile course has to be available. (It is a great deal of work to measure off a mile along a road, and do it accurately.)

In any case, the method described here can be improved by using an electronic digital tachometer that works off the ignition system. With a 1% accuracy on this instrument, the measurement is good to 2.1 mi/h. Further improvements must come in the tire size measurement—remedied somewhat if the tires are steel-belted radials, which do not distort as much as other tires at high speeds. ◄

Significant Digits

One more thing needs to be said about the propagation of uncertainty. In many instances the uncertainty needs to be known only very crudely. If we are estimating the number of gallons of paint to paint a bridge, or the number of squares of shingles to cover a roof, we are hardly justified in going through a detailed analysis of the uncertainty in the dependent variable. The same is true if we are making hypothetical design calculations. In such cases it is common to allow the uncertainty to be represented in the number of significant digits of the result.

We already saw in chapter 4 that a distance of 114.21 m is to be interpreted as 114.21 ± 0.005 m, and that a pressure of 2 400 torr should be interpreted as 2 400 ± 50 torr. In the first case we may say that the figure has five-digit accuracy, or that it has five significant digits, or that it is correct to five digits. In the second case there are only two significant digits.

Table 7-2. *Examples of Place Accuracy of Numbers*

5-digit accuracy	4-digit accuracy	3-digit accuracy	2-digit accuracy	1-digit accuracy
4.312 7	4.313	4.31	4.3	4
724.96	725.0	725	720	700
12 399	12 4$\bar{0}$0	12 400	12 000	10 000
560 0$\bar{0}$0	560 $\bar{0}$00	56$\bar{0}$ 000	560 000	600 000
0.001 213 1	0.001 213	0.001 21	0.001 2	0.001

This number of significant digits has nothing to do with the position of the decimal point. Table 7-2 gives examples. We are simply talking about the number of digits that *mean something*. In the number 725.0, the zero means that the "true value" is closer to 725 than to 725.1 or to 724.9, so we have four-digit accuracy, or four significant digits. In the number 720, the zero is simply the holder of a place value; the true value may be anywhere between 715 and 725. The number 720 therefore has two significant digits. On the other hand, the number 72$\bar{0}$ has three significant digits, since it implies that the true value is between 719 and 721.

> **DEFINITION** — The *significant digits* in a number are those that represent a measurable part of the quantity. Digits other than zero are always significant. A zero is significant only when a line has been placed over it ($\bar{0}$), when it sits between nonzero digits, or when it falls to the right of the decimal portion of a number.

With this in mind, we can formulate a crude rule of thumb for the propagation of uncertainty. This rule is useful when a detailed analysis is not justified:

> **SIGNIFICANT DIGIT RULE** — In any calculation, the number of significant digits in the result should not exceed the number of significant digits in the least accurate number used in the calculation.

Significant Digits in Calculation

Suppose we are doing a calculation with two numbers. We may be adding them, subtracting them, multiplying them, taking sines or cosines, squaring them, or doing anything to them. Let's say that one number has seven significant digits and the other has three. Then the *significant digit rule* says that the result can have no more than three significant digits. Just as a chain can be only as strong as its weakest link, a calculation can be no more accurate than its least accurate part. Let's look at another example.

Using the Significant Digit Rule

▶ **Example 7-8: Indirect measurement of impedance**
The impedance of an alternating-current (ac) electric circuit is defined as the ratio of the voltage to the current. If Z is the impedance in ohms (Ω), V is the voltage in volts, and I is the current in amperes, we have

$$Z = \frac{V}{I}$$

Suppose that the voltage and the current are measured, with the following results. What is the impedance?

$$V = 97 \text{ V}$$
$$I = 0.37 \text{ A}$$

If we do the calculation for the best value, we get

$$Z = \frac{97 \text{ V}}{0.37 \text{ A}}$$
$$Z = 262.162\ 162\ 162\ 162\ \Omega$$

According to the significant digit rule, we cannot leave the result like this: it can have no more than two significant digits. Doing this calculation, in other words, can in no way *improve* on the original measurements. What we have to be content with, then, is

$$Z = 260\ \Omega \blacktriangleleft$$

Limitations of the Significant Digit Rule

Now the significant digit rule has a slight catch, which is why it was not mentioned until this chapter. The catch is that the rule does not actually tell how many significant digits to keep in the result. In saying, "the number of significant digits in the result *should not exceed*," it tells only how many is too many.

It is not hard to come up with examples where this leads to trouble. Suppose we are calculating a quantity Q, according to

$$Q = \frac{86.7}{0.102}$$

We get a quotient of 850.000 00, which we round off by the significant digit rule to 85̄0. This three-digit accuracy corresponds to the three-digit accuracy of the original numbers. But if we look at the two numbers as

$$86.7 \pm 0.05 \quad \text{and} \quad 0.102 \pm 0.000\ 5$$

and follow either of the first two procedures in the chapter, we get

$$Q = 850 \pm 5$$

In other words, the result is actually good to only two significant digits, not to three.

The thing to remember about the significant digit rule is that it may imply that we have more accuracy than we really do. This is particularly the case with calculations involving many steps and/or many numbers. Still, we should follow this rule any time we are not going to bother with the detailed propagation of uncertainty calculations. It is much better than nothing, and most of the time it will not be wrong by more than a factor of ten.

Summary

Measurements are often made indirectly, by measuring one or more quantities that the quantity of interest depends on, then by doing a calculation. The errors in the actual measurements will always contribute to an error in the calculated quantity. The uncertainties will do likewise. We say, then, that the uncertainty *propagates*.

This will also be true in design calculations. The independent variables will be measured, or at least measurable. The uncertainty in the final result must reflect the uncertainties in these quantities.

If the dependent variable is found by reading a graph, this same graph can be used to find the upper and lower error limits. If the calculation is made by using a formula, the same formula can be used to find the upper and lower limits of the calculated quantity. The uncertainties can also be computed directly by using the formulas in table 7-1. If these techniques are not used (as in a quick calculation), at the very least the *significant digit rule* should be followed. This rule says that the result should never be quoted with more significant digits than the least accurate quantity used in the calculation.

REVIEW QUESTIONS

1. What is an indirect measurement? Give an example.

2. Why are some measurements made indirectly?

3. If two quantities are related through a graph, explain how to find the uncertainty in the dependent variable from the uncertainty in the independent variable.

4. How might the graph itself contribute to the uncertainty in an indirect measurement based on it?

5. What is the propagation of uncertainty?
6. If two quantities are related through an algebraic formula, explain how to find the uncertainty in the dependent variable from the uncertainty in the independent variable.
7. Outline the procedure for finding the uncertainty in a calculated quantity that depends on two or more measured variables.
8. What are the formulas in table 7-1 used for?
9. What are significant digits?
10. What is the significant digit rule for the propagation of uncertainty?

EXERCISES

1. A certain concrete column has an H/D ratio of 1.03 ± 0.15. Use figure 7-3 to find its strength.

2. A cylindrical concrete column is to have a strength between 155% and 165% of the strength of a standard column. What should its H/D ratio be? (*Answer:* $H/D = 0.67 \pm 0.04$)

3. The voltage produced by an iron-constantan thermocouple is 6.2 ± 1.0 mV. Use figure 6-5 to find the temperature and its uncertainty.

4. A tank is filled with methyl alcohol under a gauge pressure of 20 ± 3 lb/in^2. Use figure 6-13 to find the boiling point of the alcohol, and the uncertainty in this quantity.

5. A steel sample has a Brinell hardness number of 352 ± 10. Use figure 7-6 to find the yield strength and the uncertainty.

6. The *coefficient of friction* μ between two surfaces is defined by

$$\mu = \frac{f}{N}$$

where f is the force of friction and N is the force pressing the surfaces together, both expressed in the same unit. Measurements yield the following results:

$$f = 32.6 \pm 2.1 \text{ kgf}$$
$$N = 22.121 \pm 0.000\,5 \text{ kgf}$$

Find the uncertainty in μ:
a. by using the defining formula. (*Answer:* ± 0.096, dimensionless)
b. by using table 7-1. (*Answer:* ± 0.095)

7. Using the graph in figure 7-2 requires that the H/D ratio of a concrete column be measured. Suppose that the height is 5.231 ± 0.012 m and that the diameter is 1.798 ± 0.048 m. Find the H/D ratio:
   a. by using the defining formula. (*Answer:* $H/D = 2.909 \pm 0.084$, dimensionless)
   b. by using table 7-1.

8. The velocity of a water wave may be found by measuring the wavelength and the frequency, then using

$$v = f\lambda$$

where f is the frequency in hertz, λ is the wavelength in metres, and v is the velocity in metres per second. Suppose that the frequency is 0.32 Hz and the wavelength is 9.62 m. Find the velocity and its uncertainty:
   a. using one of the two methods for the propagation of uncertainty.
   b. using the significant digit rule.

9. The volume capacity of a cylindrical tank may be measured indirectly by measuring the inside radius and the height, then using

$$V = \pi r^2 h$$

Suppose that the radius is 4.23 ft and the height is 8.97 ft. Find the volume and the uncertainty in this quantity:
   a. by using a detailed propagation of uncertainty calculation.
   b. by using the significant digit rule.

10. One way of measuring the acceleration of gravity is to time the period of a simple pendulum of known length. The relationship is

$$g = \frac{4\pi^2 l}{t^2}$$

where g is the acceleration of gravity in units of m/s^2, l is the pendulum's length in metres, and t is the pendulum's period in seconds. Suppose that the length is measured to be $1.016\ 2 \pm 0.000\ 1$ m, and the period is $2.023\ 10 \pm 0.000\ 06$ s.
   a. What is the acceleration of gravity and its uncertainty?
   b. If the measurement was made at sea level, what is the latitude? (Interpolate from table 5-3.) (*Answer:* $40.1° \pm 3.4°$)

11. The rate at which heat energy is conducted through a wall may be found from

$$P = \frac{\sigma A\ \Delta T}{x}$$

where σ is the thermal conductivity in units of (Btu·in.)/(ft^2·h·F°), A is the area of the wall in square feet, x is the wall's thickness in inches, ΔT is the Fahrenheit temperature difference between the inside and the outside of the wall, and P is the rate of heat transfer in Btu's per hour. Suppose that a wall has a thermal conductivity of 0.897 (Btu·in.)/(ft^2·h·F°), a height of 8.3 ft, a length of 19.2 ft, a thickness of 4.87 in., and that the inside is at 68°F while the outside is at 12°F. Find the rate of heat transfer and its uncertainty:

a. using a detailed propagation of uncertainty calculation. (*Answer:* 1 644 ± 46 Btu/h)
b. using the significant digit rule.

12. Radar ranging equipment is used to find distances by measuring the time between an outgoing radar pulse and the reflected signal. The distance is then

$$x = \tfrac{1}{2}vt$$

where v is the speed of the radar pulse: $2.997\,051 \times 10^8 \pm 400$ m/s in air. Find the allowable uncertainty in the time measurement if the acceptable uncertainty in the distance is:

a. ±10%
b. ±1.0%
c. ±0.05%

CHAPTER 8

Empirical Equations

We have seen how measurement data can be displayed in tables and in graphs. In this chapter we will look at a third possibility: reducing a set of data to a single equation. Such an equation is called an *empirical equation,* which means that it is based on experiment rather than on theory.

DEFINITION | *An empirical equation is an equation that is developed to be approximately consistent with a given set of measurement data.*

Now many of the equations that we encounter are *not* empirical. The volume of a sphere, for instance, may be calculated from its radius by using the equation

$$V = \frac{4}{3}\pi r^3$$

This equation did not come about by making a bunch of measurements on basketballs or marbles. Rather, it resulted from the basic axioms of euclidean geometry, and so it is exact rather than approximate. If we measure the radius of a ball and find that the equation does not give the correct volume, then either our radius measurement is in error, or the ball is not a geometrically perfect sphere. The equation itself cannot be at fault.

Empirical Equation for the Speed of Sound

Let's suppose that we want to know how the speed of sound in air is affected by the air's temperature. We make a series of measurements, with the results listed in table 8-1. An empirical equation that fits this data is:

$$v = (331.4 + 0.59T) \text{ m/s}$$

where v is the speed of sound and T is the Celsius temperature. When $T = -12.51°C$, the equation gives $v = 324.0$ m/s, which is approximately the value in the table. When $T = 0°C$, the equation gives $v = 331.4$ m/s, which is exactly the value in the table. And so on. The equation is approximately consistent with the entire set of data.

Does an empirical equation offer any advantages over a table or graph? Yes, it does.

Advantages of Empirical Equations

1. An empirical equation takes up less space than a table or graph.
2. An empirical equation is easier to remember than a table or graph.
3. An empirical equation is often more accurate than a graph of the same data.
4. An empirical equation automatically interpolates from the data.

The first two advantages are obvious. The third one comes about because a graph's accuracy is limited by the coarseness of the pencil line and the grid it is drawn on. An empirical equation often (but not always) retains accuracy that would be lost on a graph. The fourth advantage, automatic interpolation, is the most important one, however. It means that the empirical equation can give

Table 8-1. *Results of Speed-of-Sound Measurements in Dry Air at Different Temperatures*

Temperature (°C)	Speed (m/s)
−12.51	324.2 ± 0.2
0.00	331.4 ± 0.1
8.82	336.8 ± 0.2
20.17	343.1 ± 0.2
32.63	350.3 ± 0.2

data values that were not part of the original set of measurements. We see a case of this in example 8-1.

Using an Empirical Equation

▶ **Example 8-1: Thunder and lightning and the speed of sound**
A study is done to record the places that lightning strikes during a typical summer thunderstorm. The angular position of a flash of lightning can be simply determined by visual observation and a compass. The distance from the point of observation is found by measuring the time between the lightning flash and the beginning of the resulting thunderclap. Multiplying this time by the speed of sound gives the distance:

$$d = vt$$

Before we can do this calculation, we need to know the speed of sound. Fortunately, we need not measure this quantity, since we already have an empirical equation for it. All we need to do is to measure the air temperature. Let's suppose that the temperature is $17.8°C$ $(64.0°F)$. Then, using the empirical equation for the speed of sound, we have

$$v = (331.4 + 0.59T) \text{ m/s}$$
$$= 331.4 + (0.59)(17.8)$$
$$= 341.9$$
$$v \approx 342 \text{ m/s}$$

We have rounded off this last value in accordance with the significant digit rule discussed in chapter 7.

The point here is that a temperature of $17.8°C$ was not part of the data in table 8-1. The empirical equation has therefore *interpolated* for us automatically. Certainly using this equation is a great deal easier than carrying the entire table with us and interpolating from it.

Uncertainties Still Propagate

What about the distance to the lightning flash? Using a stopwatch, we might measure a time of 4.1 s between the flash and the thunder. If so, the distance is

$$d = vt$$
$$= (342 \text{ m/s})(4.1 \text{ s})$$
$$d = 1\ 400 \text{ m (about 4 600 ft)}$$

Notice that the significant digit rule tells us to round this answer off to two-digit accuracy (the same as the time measurement). Doing so gives us a rather large uncertainty of ±50 m, but if we really wanted to hunt for where the lightning had struck, this result would take us close enough to find it.

A better approach here would be to work out the actual uncertainty with ±0.5 m/s uncertainty in the speed and ±0.05 s uncertainty in time. Using the techniques learned in chapter 7, this gives a total uncertainty of only ±19 m, so

$$d = 1\,400 \pm 19 \text{ m}$$

This result certainly would narrow down the hunt. ◄

Equation of a Straight Line

Now that we have seen how an empirical equation might be used, let's look at how we derive such an equation from a set of data. The best way to start is to graph the data. If this graph is close to being a straight line, the procedure is fairly simple.

In figure 8-1, we see a set of data and its graph. Since the graph is a straight line, we immediately know that the data satisfies an equation of the form

$$y = mx + b \qquad (8\text{-}1)$$

where y is the dependent variable, x is the independent variable, m is the *slope* of the line, and b is the *y intercept*.

DEFINITION | The *slope* of a straight line is the ratio of the rise to the run between any two points on the line. Representing the slope by the letter m,

$$m = \frac{\text{rise}}{\text{run}} \qquad (8\text{-}2)$$

Let's look at an example of a slope calculation.

Finding Slope

▶ **Example 8-2: The slope of the line in figure 8-1**
The rise and the run have already been indicated for two points on this line. Using these points, we have

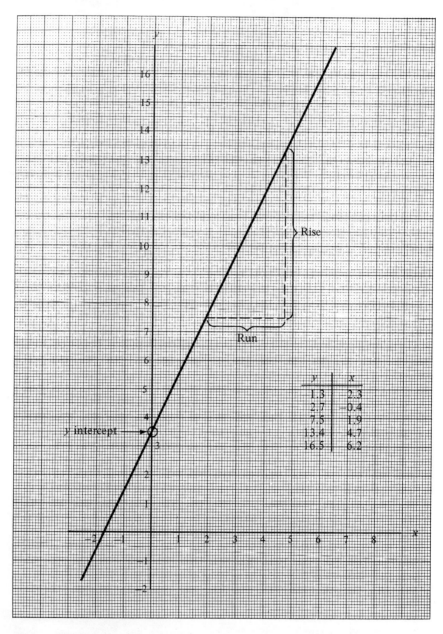

Figure 8-1. *The first step in developing an empirical equation is to graph the data.*

$$\text{rise} = 13.4 - 7.5 = 5.9$$
$$\text{run} = 4.7 - 1.9 = 2.8$$

Then the above definition of the slope gives

$$m = \frac{\text{rise}}{\text{run}}$$
$$= \frac{5.9}{2.8}$$
$$m = 2.1$$

There was nothing special about the two points we picked. Any other two points on the line will work just as well. The rise and the run might be different, but the slope will come out the same. Suppose we take the first point and the last point. Calculating the rise as the difference in *y* values and the run as the difference in *x* values, we have:

$$\text{rise} = 16.5 - (-1.3) = 17.8$$
$$\text{run} = 6.2 - (-2.3) = 8.5$$

The slope is the ratio of these, so

$$m = \frac{17.8}{8.5}$$
$$m = 2.1$$

just as before. It is a characteristic of a straight line that its slope is the same no matter which pair of points is used to calculate it. In fact, this property can be used to check how straight the line really is. ◄

Finding the y intercept

In addition to the slope, we will have to deal with the *y intercept* in writing the empirical equation.

DEFINITION | The *y intercept* of a straight line, represented by the letter *b*, is the value of the dependent variable where the line crosses the vertical axis (i.e., where the independent variable has a value of zero).

In figure 8-1, for instance, the *y* intercept is 3.5.

As mentioned earlier, a straight line is guaranteed to have an equation of the form

$$y = mx + b$$

We have just seen how to find m and b. With $m = 2.1$ and $b = 3.5$, the empirical equation describing the data in figure 8-1 is

$$y = 2.1x + 3.5$$

This means that for any value of x we choose, this equation gives the value of y corresponding to a point on the line.

Deriving Linear Empirical Equations

▶ Example 8-3: Empirical equations for the lines in figure 8-2
Looking at line I first, we identify two convenient points for finding the slope. The two points chosen here are

$$(4.0, 12.0) \text{ and } (0, 6.8)$$

where they have been written in standard ordered pair notation: (x,y). Remember that the *rise* between these points is the difference in y values, while the *run* is the difference in x values. This gives

$$\text{rise} = 12.0 - 6.8 = 5.2$$
$$\text{run} = 4.0 - 0 = 4.0$$

Then the slope is

$$m = \frac{5.2}{4.0}$$
$$m = 1.3$$

Reading the y intercept directly from the graph, we get

$$b = 6.8$$

Using these two values, we can write the equation in the form $y = mx + b$:

$$y = 1.3x + 6.8$$

This is the empirical equation for line I.
 On line II the two points

$$(0, 3.0) \text{ and } (4.0, 0)$$

are marked. We have to be careful here, since the line is sloping down rather than up. With a positive run, this line has a negative rise:

$$\text{rise} = 0 - 3.0 = -3.0$$
$$\text{run} = 4.0 - 0 = 4.0$$

The slope, then, is negative:

$$m = \frac{-3.0}{4.0}$$
$$m = -0.75$$

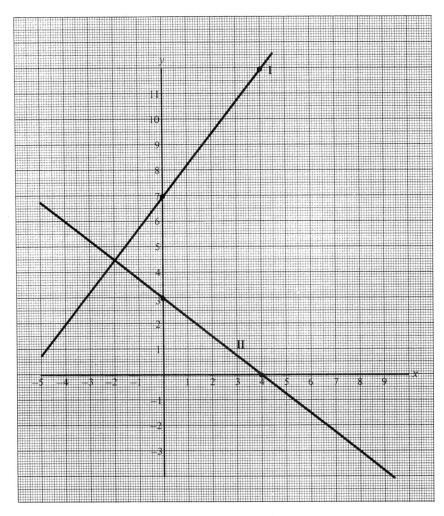

Figure 8-2. *Two straight-line graphs. The empirical equations for these lines may be written by finding the slopes and the y intercepts.*

A line that runs downward and to the right always has a negative slope. Reading the y intercept from the graph, we have

$$b = 3.0$$

The empirical equation for line II is then

$$y = -0.75x + 3.0 \quad \blacktriangleleft$$

Using the Data Itself Rather Than Its Graph

A graph is useful for visualizing the relationship between two variables. Unfortunately, as we saw in chapter 6, a graph tends to sacrifice accuracy. Thus if we want to get the best possible empirical equation that fits the data, we should really work with the data itself. There are several ways of doing so, and unfortunately most of them get fairly complicated. The simplest method that gives decent results is the *method of absolute differences*.* In fact, this method is not restricted to very accurate data, and it can be very useful for choosing the *best* straight line that fits data with large uncertainties.

Figure 8-3 shows a graph of a set of data with large uncertainties. We have reason to believe that a straight line is the best description for this data (it is, after all, the simplest curve consistent with the data). Here the problem is *which* straight line to draw. As we see from the graph, there are several possible lines that intersect the uncertainty range of each data point. If we allow near misses, there are even more. The method of absolute differences gives us the empirical equation for the single line that *best* fits this data. This method is based on choosing an average line that goes through the *center of gravity* of the data.

> **DEFINITION** | The <u>center of gravity</u> of a set of data is a point whose y coordinate is the average of all the y values in the data, and whose x coordinate is the average of all the x values in the data. If the number of data points is N, then

$$x_{cg} = \frac{1}{N} \sum_{i=1}^{N} x_i \qquad (8\text{-}3)$$

$$y_{cg} = \frac{1}{N} \sum_{i=1}^{N} y_i \qquad (8\text{-}4)$$

This definition's use is illustrated in example 8-4.

Finding the Center of Gravity

▶ **Example 8-4: Center of gravity of the data in figure 8-3**

The center of gravity is calculated through a direct application of equa-

---

*You may also hear about the *method of least squares*. That method has the drawback of being computationally more complicated, and of giving two possible empirical equations, depending on which variable is considered dependent. The method of absolute differences gives only one empirical equation, which describes a line approximately midway between the two least squares lines.

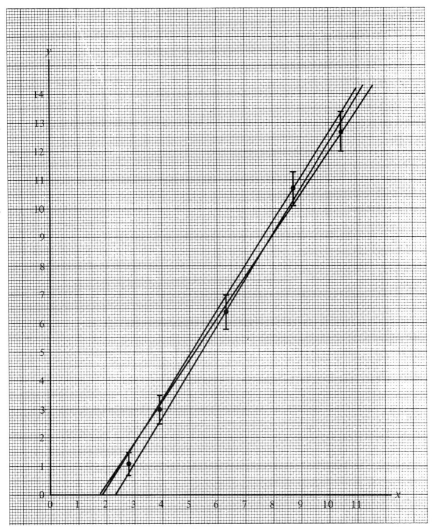

y	x
1.1 ± 0.4	2.8
3.0 ± 0.5	3.9
6.4 ± 0.6	6.3
10.7 ± 0.6	8.7
12.7 ± 0.7	10.4

Figure 8-3. *Data with large uncertainties may be consistent with more than one straight line. The method of absolute differences may be used to write the empirical equation for the "best" line.*

tions (8-3) and (8-4). Dealing with the x values of the data first, we have:

$$x_{cg} = \frac{1}{N} \sum_{i=1}^{N} x_i \qquad (8\text{-}3)$$

$$= \frac{1}{5}(2.8 + 3.9 + 6.3 + 8.7 + 10.4)$$

$$x_{cg} = 6.42$$

The value of y_{cg} is calculated similarly:

$$y_{cg} = \frac{1}{N} \sum_{i=1}^{N} y_i \qquad (8\text{-}4)$$

$$= \frac{1}{5}(1.1 + 3.0 + 6.4 + 10.7 + 12.7)$$

$$y_{cg} = 6.78$$

The center of gravity, then, is the point (6.42, 6.78). ◀

Using this idea of center of gravity, we may describe the method of absolute differences as follows:

Method of Absolute Differences for Finding the Equation of the Best Straight Line

1. Calculate the center of gravity of the data. This gives two numbers: y_{cg} and x_{cg}.
2. Calculate the differences between each x value and x_{cg}, and the difference between each y value and y_{cg}. It is usually best to keep track of these differences in a working table.
3. Take the absolute value of each of these differences. In other words, some of these differences will be negative, and we discard the negative signs. If there are N data points, this gives us N absolute differences in x and N more in y.
4. Add all the absolute differences in x, and separately add all the absolute differences in y.
5. The ratio of the sum of the absolute differences in y to the sum of the absolute differences in x gives us the slope of the line. In symbols,

$$m = \frac{\sum_{i=1}^{N} |y_i - y_{cg}|}{\sum_{i=1}^{N} |x_i - x_{cg}|} \qquad (8\text{-}5)$$

6. The y intercept, b, is found from the formula

$$b = y_{cg} - mx_{cg} \qquad (8\text{-}6)$$

In other words, we multiply the slope from step 5 by the x center of gravity, then subtract this from the y center of gravity.

7. The empirical equation *best* fitting the data is again

$$y = mx + b$$

where m is calculated in step 5 and b is found in step 6.

If this seems like a lot of work, it is. Fortunately, however, it is more time consuming than difficult, since all the steps are simple arithmetic. Let's look at an example.

Using the Method of Absolute Differences

▶ **Example 8-5: Empirical equation for the data in figure 8-3**
The first step is to find the center of gravity of the data. We already did this in the last example, with the result

$$x_{cg} = 6.42$$
$$y_{cg} = 6.78$$

We now use these values to calculate a set of absolute differences from the x and y data. Listing these differences in a working table, we have:

y_i	absolute difference from y_{cg}	x_i	absolute difference from x_{cg}
1.1	5.32	2.8	3.98
3.0	3.42	3.9	2.88
6.4	0.02	6.3	0.48
10.7	4.28	8.7	1.92
12.7	6.28	10.4	3.62

The next step is to sum these absolute differences:

$$\sum_{i=1}^{5} |y_i - y_{cg}| = 19.32$$

$$\sum_{i=1}^{5} |x_i - x_{cg}| = 12.88$$

The slope of our line is calculated from equation (8–5), which tells us to take the ratio of these two numbers.

$$m = \frac{19.32}{12.88}$$

$$m = 1.50$$

We now use equation (8–6) to get the y intercept:

$$b = y_{cg} - m x_{cg}$$
$$= 6.78 - (1.50)(6.42)$$
$$b = -2.85$$

This negative value for the y intercept should not be surprising, since any of the lines in figure 8–3, if extended, would cross the negative part of the y axis.

Our result for the empirical equation is then

$$y = mx + b$$
$$y = 1.50x - 2.85$$

Accuracy of the Empirical Equation

One final point should be made here. Although we have retained three-digit accuracy for the constants in the empirical equation, the equation itself is by no means this accurate. Since we based this equation on data where the dependent variable had an uncertainty of at least ±0.4, this is also the minimum uncertainty in any calculation done with the final equation. Moreover, if we use values of x that are much greater than 10.4 or much less than 2.8, the equation will be considerably less accurate than this. Why? Because the use of values beyond the original range in x would constitute an *extrapolation* rather than an interpolation. And as we saw in chapter 5, extrapolations are always unreliable. ◄

When the y intercept Is Zero

We sometimes encounter situations where we know before we start that the y intercept, b, has to be zero. If an engine's power output depends on its fuel consumption, then we know that the power output will be zero when no fuel is consumed. This has nothing to do with mathematics—it is merely a property of engines. If we are looking at the relationship between the powder charge in a cartridge and the muzzle velocity of the bullet, we automatically know that there is no muzzle velocity when there is no charge.

Again, this is based on what we know about ammunition. Empirical equations that govern cases like this will have a value of $b = 0$. If the graph of such data is a straight line, the empirical equation then reduces to

$$y = mx \qquad (8\text{-}7)$$

This can make things very simple. We either measure the slope m from the graph or calculate it from a simplified form of the method of absolute differences,

$$m = \frac{y_{cg}}{x_{cg}} \qquad (8\text{-}8)$$

Let's look at an example.

Using the Method of Absolute Differences When y intercept Is Zero

▶ **Example 8-6: Friction**
The force of friction between a pair of surfaces depends on the force pressing the surfaces together. This latter force is called the *normal* force. Below is some data for a piece of brake lining in contact with cast iron. Let's find the empirical equation describing this data.

Force of Friction (N)	Normal Force (N)
4.2 ± 1.5	12.1
16.8 ± 3.0	38.4
26.1 ± 5.8	73.2
48.2 ± 6.0	108
56.7 ± 6.2	142
64.2 ± 7.9	178

One thing we know about friction is that there is none if the surfaces are not pressed together. The best straight line fitting the data must therefore go through the point (0,0), as shown in figure 8-4. Because the y intercept is zero, the empirical equation must have the form

$$y = mx \qquad (8\text{-}7)$$

or, using more appropriate symbols,

$$f = mn$$

where f is the force of friction and n is the normal force. As we have just seen, the constant m (which is still the slope of the line) can be calculated fairly simply. First we find the center of gravity of the data:

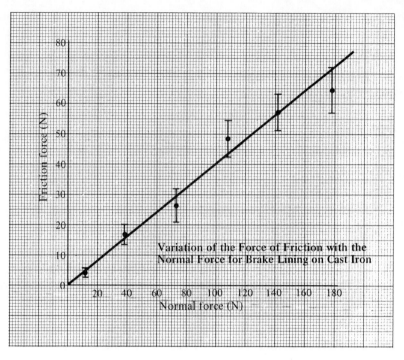

Figure 8-4. *Graph for the empirical equation in example 8-6.*

$$f_{cg} = \frac{1}{N} \sum_{i=1}^{N} f_i$$

$$= \frac{1}{6}(4.2 + 16.8 + 26.1 + 48.2 + 56.7 + 64.2)$$

$$= \frac{1}{6}(216.2)$$

so $\quad f_{cg} = 36.0$

and $\quad n_{cg} = \frac{1}{N} \sum_{i=1}^{N} n_i$

$$= \frac{1}{6}(12.1 + 38.4 + 73.2 + 108 + 142 + 178)$$

$$= \frac{1}{6}(551.7)$$

so $\quad n_{cg} = 92.0$

Then, from equation (8-8), the slope is

$$m = \frac{f_{cg}}{n_{cg}}$$

$$= \frac{36.0}{92.0}$$

$$m = 0.39$$

This gives an empirical equation of

$$f = 0.39n$$

Coefficient of Friction

This coefficient 0.39 is called the *coefficient of friction*. It is numerically equal to the average slope of a graph of f versus n. Values of the coefficients of friction for various pairs of surfaces in contact may be found in most engineering and machinist's handbooks. Most such values have been determined by an analysis similar to the one described here. We might also note that large uncertainties are always characteristic of friction measurements. ◄

Let's summarize what we have done so far. If the graph of a set of data is a straight line, then we know that the empirical equation is $y = mx + b$. The problem is to find the best numerical values for m and b. There are two possible approaches: (1) If we don't need extreme accuracy, we can read m and b directly from the graph; (2) if we need the *best* equation consistent with the data, we can use the method of absolute differences. Either way, the equation is still $y = mx + b$, with the possibility that b might be zero.

Nonlinear Data: Power from the Wind

We now need to consider the case of graphs that are *not* straight lines. In table 8-2, we see some data on the gross power output of a small wind generator at different wind velocities. (A wind*mill* is something that grinds grain; a wind *generator* produces electricity.) This power may be measured in a number of ways. One way is to connect an electrical load to the generator, then measure the current and the voltage. If the generator itself does not have a large electrical resistance, the power may be calculated as the product of these two measurements. Another method to measure the power is to connect the generator to a resistance heater that is immersed in a measured quantity of water in an insu-

Table 8-2. *Data on the Gross Power Output of a Small Wind Generator with a Blade Diameter of About 3 m*

Power (W)	Wind Speed (km/h)
0	0
9.6	8.6
76.5	17.2
378	29.3
69$\bar{0}$	35.8
1 380	45.1

lated container. The rate at which the water temperature rises can be used to calculate the generator's power output. There are some other, slightly more complicated methods, but regardless of the method used, the wind speed must be carefully controlled during the measurement. Thus it is best to make the measurements in a wind tunnel.

In any case, if we graph the data in table 8-2, we most definitely do not get a straight line. Without a straight line, we cannot write the empirical equation. Yet if we are told that the empirical equation is

$$P = 0.015\,0 v^3 \text{ W}$$

where v is the wind speed in km/h, we can easily check it. Substituting a wind speed of 17.2 km/h gives

$$P = (0.015\,0)(17.2)^3 \text{ W}$$

$$P = 76.3 \text{ W}$$

which is very close to the data value 76.5 in table 8-2. You can verify that the other values in the table are also consistent with this empirical equation.

Equations for Curves

But we *can* develop the equation ourselves. First we graph the data, and find that it is not linear (see figure 8-5a). The problem is that the line curves upward too steeply. So we *guess* some other possible formulas and try them out. Perhaps

$$P = kv^2$$

▶ **Figure 8-5.** *Working graphs used to find the empirical equation for the set of nonlinear data in tables 8-2 and 8-3. (a) Graph of P versus v; (b) graph of P versus v^2; (c) graph of P versus v^3.*

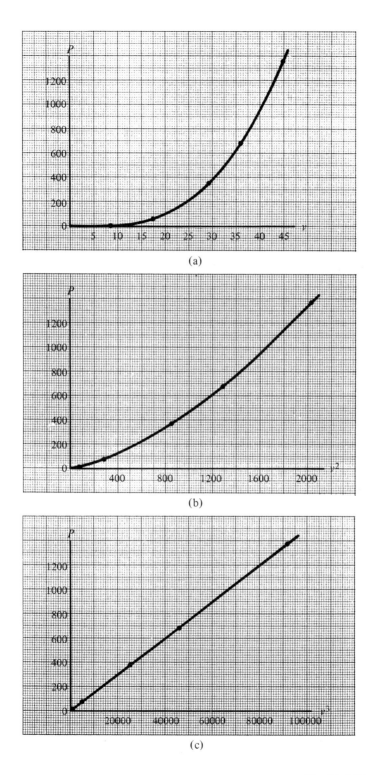

(a)

(b)

(c)

Table 8-3. *A Working Table for Determining Whether the Power of a Wind Generator Is Proportional to v^2 or to v^3*

P	v	v^2	v^3
0	0	0	0
9.6	8.6	74.0	636
76.5	17.2	296	5 090
378	29.3	858	25 200
690	35.8	1 280	45 900
1 380	45.1	2 030	91 700

Note: The values of v^2 and v^3 are calculated from the measured values of v.

where k is a numerical constant we can figure out later. Or perhaps

$$P = kv^3$$

These equations would make P increase very rapidly with v, just as it does. To try them out, we calculate the values of v^2 and v^3 from the original data. The results are listed in table 8-3. Then we graph P versus the calculated values of v^2. If this is a straight line, then $P = kv^2$ is the correct empirical equation. Unfortunately, figure 8-5b shows that it is not. So we graph P versus v^3. This graph is shown in figure 8-5c, and it *is* a straight line. So the correct empirical equation is

$$P = kv^3$$

How do we find the constant k? The same as before. Since $P = kv^3$ is a straight line, it must be the same as

$$y = mx + b$$

Instead of y, we have P. Instead of m, we have k. Instead of x, we have v^3. The constant b is zero, since there is no power when there is no wind. The constant k we are looking for is just the slope of the line. We can verify that this slope is 0.015 0, which gives the correct empirical equation.

The Intelligent Guess

An important part of this procedure is the guess. To make an intelligent first guess, it is important to be familiar with the general shapes of the graphs of common algebraic equations. Figure 8-6 shows a number of such graphs. Because we seldom encounter negative values of an independent variable, the left-hand portions of these curves are shown as dashed lines. The problem is usually to fit the right-hand portion of one of these curves to a set of data. Let's look at an example.

Deriving Nonlinear Empirical Equations

▶ **Example 8-7: Some nonlinear data**
The problem is to find the empirical equation that best fits the following data:

y	x
-1.59	0.48
0.08	0.87
4.16	1.43
9.37	1.92
12.61	2.17

Our first step is to graph the data to see if it already lies on a straight line. Unfortunately, as we see in figure 8-7a, it does not. Nevertheless, this graph does give us a clue about what the right equation might be. If we compare its shape with the graphs in figure 8-6, we see that a reasonable guess is

$$y = mx^2 + b$$

To check this guess, we go back to the original data and calculate the values of x^2. Putting these values into a working table, we have

y	x	x^2
-1.59	0.48	0.230
0.08	0.87	0.757
4.16	1.43	2.045
9.37	1.92	3.686
12.61	2.17	4.709

We now draw another graph (figure 8-7b), using the values of y versus x^2 from this table. Since the data points do fall on a straight line, we conclude that

$$y = mx^2 + b$$

is a valid empirical equation for the data.
The only problem that remains is the determination of m and b. If we are not concerned about extreme accuracy, we may find them directly from the graph. Reading the value of the y intercept, we get

$$b = -2.3$$

For the slope, we can use any two points on the line. The procedure is somewhat more accurate, though, if we pick points that are far apart. If we use the first and the last data points, we get

Empirical Equations / 139

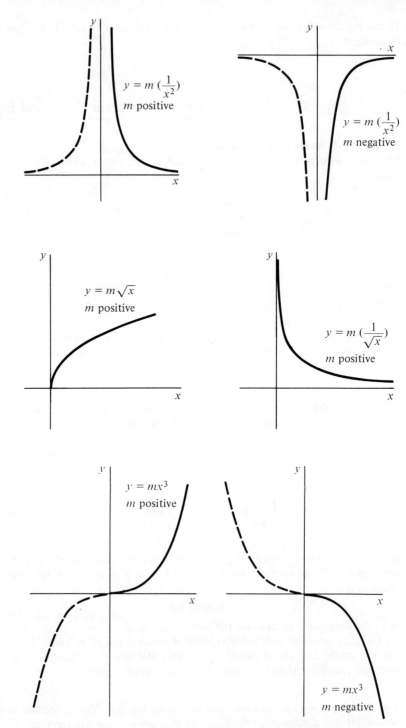

Figure 8-6. *Graphs of some common algebraic functions.*

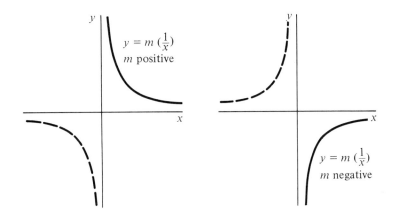

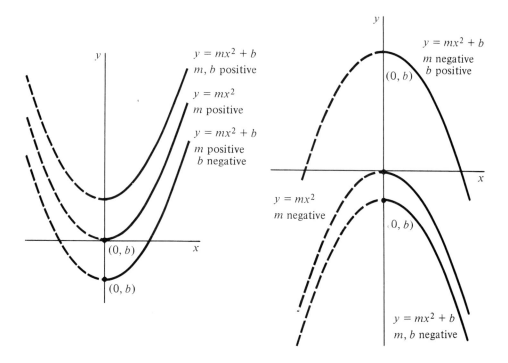

Figure 8-6. *(Continued)*

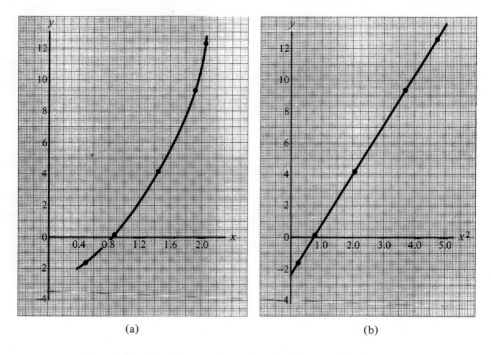

Figure 8-7. *Working graphs used to find the empirical equation for the data in example 8-7. (a) Graph of y versus x; (b) graph of y versus x^2.*

$$m = \frac{\text{rise}}{\text{run}} \tag{8-2}$$

$$= \frac{12.61 - (-1.59)}{4.709 - 0.230}$$

$$= \frac{14.20}{4.479}$$

$$m = 3.17$$

The empirical equation is then

$$y = 3.17x^2 - 2.3 \blacktriangleleft$$

Finally, let's examine a case of the same analysis applied to a set of precision measurement data.

Absolute Differences and Nonlinear Data

▶ **Example 8-8: Dispersion curve of a sample of glass**
Glass and other optical materials have the ability to bend light that passes through them. The extent of this bending is characterized by a quantity known as the *index of refraction*. This quantity, symbolized n, is defined as

$$n = \frac{\text{speed of light in substance}}{\text{speed of light in vacuum}}$$

In practice, the index of refraction is measured indirectly by using a *spectrometer*.

The index of refraction is slightly different for different colors of light, which is why a glass prism splits white light into the colors of the rainbow. It is also why cheaply made optical instruments do the same thing. (This defect is called *chromatic aberration*.) Suppose we need to know exactly how the index of refraction of a certain glass sample depends on the color of the light, information that can be very important in optical design work. We can quantify color by measuring the light's wavelength. The approximate relationship between color and wavelength is shown in table 8-4.

Table 8-4. *Approximate Wavelengths of Various Colors of Light*

Color	Wavelength (Å)
ultraviolet (not visible)	less than 4 000
violet	4 100
indigo	4 500
blue	4 800
green	5 400
yellow	5 700
orange	6 300
red	6 800
infrared (not visible)	greater than 7 000

Wavelength Standards

Most light is composed of a mixture of many, many wavelengths, even if it looks like it has just one color. To do an experiment like this, we need

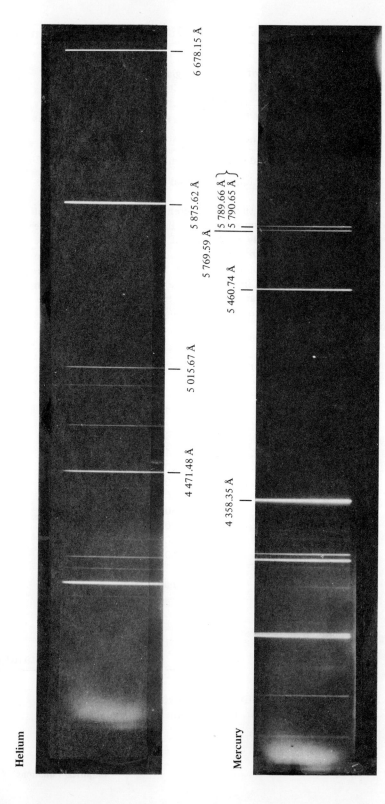

Figure 8–8. Line spectra from helium and mercury gas discharges. The emission lines may be used as laboratory standards for wavelength.

to have several sources of light with fairly well-defined wavelengths. One practical source is a gas-discharge light source. Light from such a source has only a small number of well-defined wavelengths, and the unwanted wavelengths are easily filtered out. Figure 8-8 shows photographs of the spectra from a mercury and a helium gas discharge as they appear in a spectrometer. We have identified the wavelengths of some of the spectral lines on these photographs. These wavelengths have been measured very accurately and tabulated in a number of sources; the Chemical Rubber Company's *Handbook of Chemistry and Physics* is one. Gas-discharge spectral lines are very good laboratory standards for wavelength, since they have been measured by direct comparison with the SI primary standard of length.

Now suppose that we project some of these wavelengths through our sample and measure the index of refraction for each one. The results are given in table 8-5. Based on this data, what is the empirical equation relating the index of refraction to the wavelength of the light?

Now, a graph of this data is not a straight line, as we see in figure 8-9a. Consulting figure 8-6, we see that a reasonable guess is that the index of refraction n depends on $1/\lambda^2$, where λ is the wavelength.

To test this guess, we make up a working table (table 8-6) to calculate the values of $1/\lambda^2$ from the original data. A graph of this data (figure 8-9b) is a straight line, which tells us that the empirical equation is of the form

$$n = m \frac{1}{\lambda^2} + b$$

The problem then reduces to finding m and b. Notice that we cannot read b directly from the graph since the horizontal axis does not start at zero. Rather than draw the graph again, let's use the method of absolute differences, which is more accurate anyway.

There are seven data points. The center of gravity of n is 1.649 07, while the center of gravity of $1/\lambda^2$ comes out as $3.675\,34 \times 10^{-8}$. Omitting the details of the calculation, the sum of the absolute differences between the values of n and n_{cg} works out as

$$\sum_{i=1}^{7} |n_i - n_{cg}| = 0.038\,77$$

The corresponding quantity for $1/\lambda^2$ is

$$\sum_{i=1}^{7} \left| \left(\frac{1}{\lambda^2}\right)_i - \left(\frac{1}{\lambda^2}\right)_{cg} \right| = 6.429\,95 \times 10^{-8}$$

This gives us a slope of

$$m = \frac{0.038\,77}{6.429\,95} \times 10^{-8}$$

$$m = 6.029\,6 \times 10^5$$

Table 8-5. *Data on the Index of Refraction of a Sample of Glass for Light of Different Wavelengths*

Color	Source	Wavelength (Å)	Index of Refraction
red	helium	6 678.15	1.640 8 ± 0.000 5
yellow	helium	5 875.62	1.644 9 "
yellow	mercury	5 779 ± 10	1.644 6 "
green	mercury	5 460.74	1.646 6 "
green	helium	5 015.67	1.650 9 "
indigo	helium	4 471.48	1.656 7 "
violet	mercury	4 358.35	1.659 0 "

Note: The uncertainty in the yellow line of mercury comes about because this is really a cluster of three wavelengths, and no single one can be filtered out. The index of refraction has no units, since it is a ratio of two speeds.

Table 8-6. *Working Table for Testing Whether the Index of Refraction in Table 8-5 Depends on $1/\lambda^2$*

n	λ	$\dfrac{1}{\lambda^2}$
1.640 8	6 678.15	$2.242\ 27 \times 10^{-8}$
1.644 9	5 875.62	$2.896\ 63 \times 10^{-8}$
1.644 6	5 779	2.994×10^{-8}
1.646 6	5 460.74	$3.353\ 49 \times 10^{-8}$
1.650 9	5 015.67	$3.975\ 05 \times 10^{-8}$
1.656 7	4 471.48	$5.001\ 47 \times 10^{-8}$
1.659 0	4.358.35	$5.264\ 48 \times 10^{-8}$

Note: λ is used to represent the wavelength.

We find the quantity b from

$$b = n_{cg} - m\left(\frac{1}{\lambda^2}\right)_{cg}$$

so $\quad b = 1.649\ 07 - (6.029\ 6 \times 10^5)(3.675\ 34 \times 10^{-8})$

$b = 1.626\ 91$

With m and b determined, we may now write the empirical equation

$$n = m\left(\frac{1}{\lambda^2}\right) + b$$

$$n = (6.029\ 6 \times 10^5)\left(\frac{1}{\lambda^2}\right) + 1.626\ 91$$

where λ is in units of angstroms.

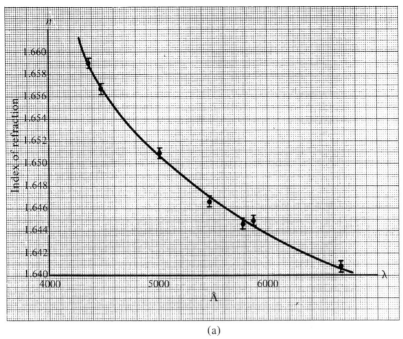

(a)

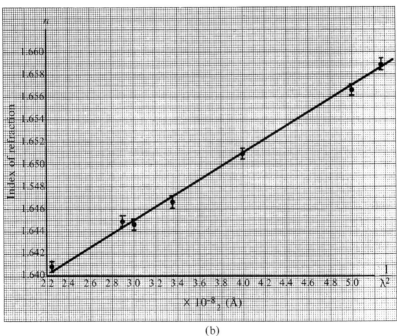

(b)

Figure 8-9. *Working graphs for determining the empirical equation for the dispersion curve of a sample of optical glass. (a) Graph of n versus λ; (b) graph of n versus $1/\lambda^2$.*

The reader should verify that this equation indeed fits the original data. Substituting any of the original values of λ into this equation gives the corresponding value of n to within the original uncertainty of $\pm 0.000\,5$. ◀

The most difficult step in finding the empirical equation is in guessing an appropriate trial function. After this is found, the rest of the analysis is straightforward. In chapter 10, we will look at a procedure that often simplifies this first step of finding the form of the equation.

Summary

Most equations encountered in the sciences are definitions of quantities, or equations derived mathematically from such definitions. Occasionally, however, we use equations derived directly from a set of measurements. Such relationships are called *empirical equations*.

Although any set of data may be represented in table or graph form, condensing it into an empirical equation offers certain advantages. The most important of these advantages is the accuracy and simplicity of interpolations from the data.

The procedure for finding an empirical equation is based on the mathematics of straight lines. If the data graph is already a straight line, the empirical equation is written directly in terms of the *slope* and the *y intercept*. If such a graph is not a straight line, the independent variable is replaced by a trial function of that variable, a new graph is drawn with the trial function as the independent variable, and the process is repeated until a trial function is found that yields a straight-line graph. Again, the equation is written in terms of the slope and the *y* intercept. For cases where a graphical determination of the slope will not give the desired accuracy, the *method of absolute differences* may be used. This requires a calculation of the *center of gravity* of the data. The slope and the *y* intercept are then calculated from this quantity.

REVIEW QUESTIONS

1. What is an empirical equation?

2. Give an example of an equation that is not empirical.

3. What advantage does an empirical equation offer over a table or graph of the same data?

4. What is the empirical equation describing data that has a straight-line graph?

5. What is the slope of a straight line? How is it found?

6. What is the *y* intercept of a straight line?

7. Outline the procedure for writing the empirical equation of data that has a straight-line graph.

8. What is the method of absolute differences used for?

9. What is meant by the center of gravity of a set of data?

10. Outline the procedure for calculating the center of gravity of a set of data.

11. Outline the method of absolute differences.

12. How is the method of absolute differences simplified if the data includes the point (0,0)?

13. What is the basic procedure for finding the empirical equation if the data graph is not a straight line?

EXERCISES

1. Graph the straight lines having the following slopes and *y* intercepts:
   a. $m = 3, b = 0$
   b. $m = -3, b = 0$
   c. $m = 0, b = 4$
   d. $m = 2, b = -3$
   e. $m = -0.5, b = 4$

2. Write the empirical equations of the graphs in exercise 1. [*Answer:* (d) $y = 2x - 3$]

3. Find the empirical equation for the following set of data by reading the slope and *y* intercept from a graph.

y	x
0.5	0.9
2.5	1.8
3.9	2.4
6.9	3.7

   (*Answer:* $y = 2.3x - 1.6$)

4. Find the slope and *y* intercept of the following data and write the empirical equation.

y	x
3.15	1.03
3.51	1.86
3.79	2.51
4.31	3.73
4.82	4.90

5. Find the empirical equations by looking for a relationship that yields a straight-line graph, then by reading the slope and y intercept from the graph.

a.

y	x
3.66	0.60
3.08	1.25
-2.67	7.63
-5.16	10.4
-7.14	12.6

b.

y	x
12.0	0.25
3.26	0.92
2.54	1.18
1.11	2.71
0.38	7.82

c.

y	x
0	1.00
0.68	1.52
1.05	2.11
1.30	2.86
1.58	4.77
1.80	10.1
1.98	96.2

(Answer: $y = 2.0 - \dfrac{2.0}{x}$)

d.

y	x
9.26	2.1
14.20	2.6
16.46	2.8
18.90	3.0
24.28	3.4
25.73	3.5
35.30	4.1

e.

y	x
0.027	0.10
0.422	0.25
1.16	0.35
3.38	0.50
13.8	0.80

f.

y	x
430 ± 5	10.3
30 600 ± 200	86.4
60 100 ± 300	121
151 000 ± 500	192
167 000 ± 500	202

(Answer: $y = 4.1x^2$)

g.

y	x
0	0
4.49	2.1
6.65	4.6
9.75	9.9
11.1	12.7

h.

y	x
2.0	0
1.0	1.0
0.36	2.7
-0.93	8.6
-2.05	16.4

(Answer: $y = 2.0 - \sqrt{x}$)

i.

y	x
2.92	1.2
1.71	3.5
1.16	7.6
0.91	12.5
0.69	21.2

6. Find the empirical equation for the data in exercise 3 by using the method of absolute differences. (*Answer:* $y = 2.29x - 1.59$)

7. Find the empirical equation for the data in exercise 4 by using the method of absolute differences.

8. For each of the sets of data in exercise 5, use the method of absolute differences to find the best numerical constants in the empirical equation. [*Answers:* (d) $y = 2.10x^2$; (f) $y = 4.10x^2$; (h) $y = 2.00 - 1.00\sqrt{x}$; (i) $y = 3.199/\sqrt{x}$]

CHAPTER 9

Polar Graphs

The Cartesian Coordinate System

In chapter 6, we looked in some detail at the procedure for representing data on a graph. In every example we drew the graph on a grid of perpendicular, equally spaced lines. Such a grid is often called a *Cartesian coordinate system,* named after René Descartes, the mathematician who invented it.

DEFINITION | *A Cartesian coordinate system is a grid of perpendicular straight lines, usually equally spaced, which is used to locate points on a graph. A point is located by specifying two numbers: one describing the point's horizontal position on the grid, and one describing its vertical position. These two numbers are called the point's Cartesian coordinates.*

Notice that the preceding definition does not describe anything new; it simply gives a name to what we have been doing already.

The Cartesian coordinate system is by no means the only way of locating a point on a plane. Mathematicians have devised several dozen other systems that will do the same thing. The reason that we do not see many of these different systems is that most of them are useful only in very special situations.

The Polar Coordinate System

One non-Cartesian system does see a fair amount of use, however: the *polar coordinate system.*

DEFINITION | A _polar coordinate system_ is a grid composed of a series of concentric circles,* and radial spokes cutting through these circles. A point is located on this grid by specifying two numbers—one giving the radial distance from the center of the system (i.e., the distance along a spoke), and the other giving the angular direction of the point as measured from the center. These two numbers are called the point's _polar coordinates_.

Plotting Data on Polar Paper

A piece of polar coordinate graph paper is shown in figure 9-1. Such paper is printed by a number of different manufacturers, and is available in most office-supply and art stores. We locate points on this paper by specifying two numbers: r, the radial distance from the center, and θ (the Greek letter theta), the angular direction of the point. The conventional way of reporting these coordinates is to give r first, then θ. The coordinates of a point P, then, can be written in the form

$$P = (r, \theta)$$

For instance, the point P_1 is 4.0 units from the center on the spoke labeled 10°. We may specify the point, then, by

$$P_1 = (4, 10°)$$

Similarly, P_2 is 5.4 units from the center with a direction of 137°. Its coordinates, then, are

$$P_2 = (5.4, 137°)$$

If we are given the coordinates, it is easy to plot the point. Suppose we are to plot the point given by $P_3 = (3.6, 245°)$. We may begin by finding the spoke labeled 245°. Then, starting at the center, we travel out along this spoke a total of 3.6 units. This point P_3 has been indicated on figure 9-1.

Polar graphs are very useful for describing quantities that depend on direction. A floodlight, for instance, is not equally bright in all directions. Suppose that we need to know how the luminous intensity (the "brightness") of such a lamp depends on the angle of view. We can use 0° to represent the direction the lamp is pointing. Directly behind the lamp, then, is 180°. Here we expect to find no light at all. If we make detailed measurements of the luminous intensity at different angles from the axis of the lamp and plot the results on a polar graph, we will get a result much like figure 9-2. One interesting thing about this result

---

*Circles are _concentric_ if they have the same center. If they have different centers, we say they are _eccentric_(i.e., off center).

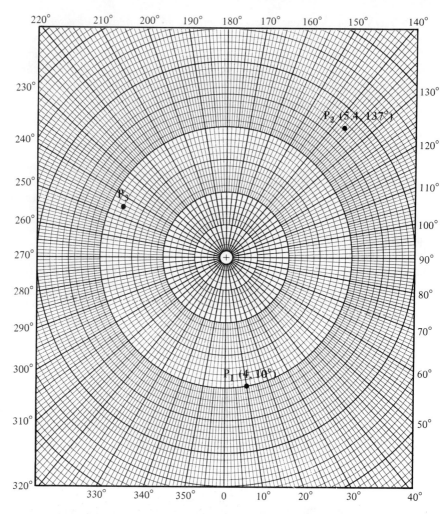

Figure 9-1. *Plotting points on polar coordinate graph paper.*

is that this particular lamp throws its brightest spot 25° from the direction it is aimed in.

Let's now examine the entire procedure for drawing a polar graph.

Drawing a Polar Graph

▶ **Example 9-1: Torque curve for a spiral spring**
Measurements are made on the torque needed to wind a flat spiral spring through a given angle. This torque, it turns out, is also the countertorque exerted by the spring. Represent the following data on a polar graph.

154 / Chapter 9

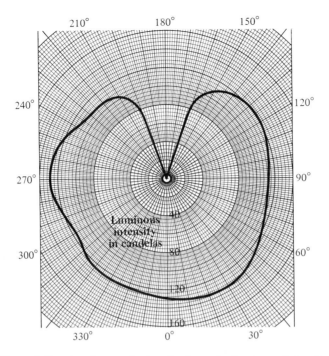

Figure 9-2. *A polar graph showing how the luminous intensity of a floodlight varies with the viewing angle.*

Torque (lb·in)	Angle (deg)
0	0
0.24	30
1.42	100
3.97	200
6.52	300
9.07	400
11.58	500
14.24	600
16.68	700
19.42	800

The independent variable here is the angle. The total range of this variable is 800°. Since polar paper indicates angles only up to 360°, it may seem that we cannot plot this data on a polar graph.

If we think about what the data means, however, we see that we really do not have any problem using polar coordinates. Winding the spring through one revolution accounts for 360° of the independent variable; this corresponds to once around a polar graph. Further winding corresponds to beginning a second revolution, both on the spring and on the graph: 400° is plotted as 400° - 360°, or 40°; 500° as 500° - 360°, or 140°; and so on. Since the data actually represents slightly more than two com-

Polar Graphs / 155

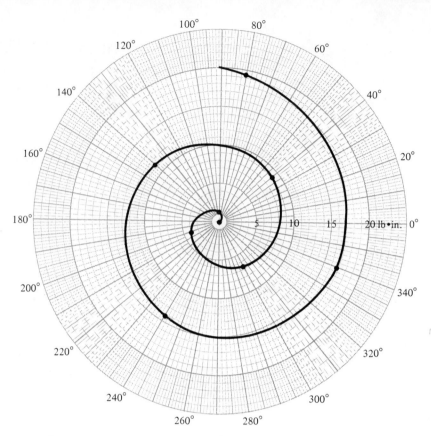

Figure 9-3. *Polar graph showing a torque curve for a flat spiral spring.*

plete revolutions of the spring, the graph will show a curve that travels through more than two revolutions.

The dependent variable, the torque, will be plotted as the radial coordinate. The range of this variable is 19.42 lb·in. Suppose we use a piece of paper with 25 radial divisions. A suitable scale, then, is 1 division = 1.0 lb·in.

Once we have decided on the radial scale, we can plot the points. Interpolating with a french curve, we get the result shown in figure 9-3. It is an interesting coincidence in this example that the torque curve has the same shape as the spiral spring itself. ◄

How do we decide whether to plot data on a polar graph? It is largely a matter of judgment, because any data that can be plotted on a Cartesian coordinate system can also be plotted on a polar coordinate system. The resulting graphs may not look very much alike, but they will still represent the same data. If, however, the data contains negative values, the resulting polar graph may become somewhat confusing, both to plot and to read. Let's restrict our attention to cases where both variables are always positive.

Comparison of Polar and Cartesian Graphs

One such case is the voltage output of an automobile alternator, shown on a Cartesian graph in figure 9-4a. This voltage is found to fluctuate as the alternator rotates, peaking out twice during each revolution of the alternator's shaft. Since the alternator is typically rotating at several hundred revolutions per minute, the voltage measurements here had to be made automatically, perhaps with an oscilloscope. Regardless, the graph is still the result of measurement. (It may seem curious that the peak voltage is as high as 25 V when the car probably has a 12-V electrical system. Actually, the *average* voltage here is closer to 16 V, just enough to charge a 12-V battery efficiently.)

The relationship between voltage and shaft orientation also can be plotted on a polar graph (figure 9-4b). If you examine these graphs carefully, you will see that they do in fact represent the same data.

Direction of Increasing Angular Coordinates on Polar Graphs

There is no definite rule on where the 0° spoke should be on a polar graph. In figure 9-2, this spoke points down, while in figure 9-3 it was off to the right. In both cases, however, the angle increased as we went around the graph in a *counterclockwise* direction.

Compass Direction

When the angle is used to indicate *compass direction,* however, standard procedure is to do just the opposite. In other words, compass headings increase as we go around the compass *clockwise.* Figure 9-5 shows the relationship between the standard compass directions and the headings in degrees.

Suppose, for instance, that we need to know how the strength of a certain radio signal depends on the direction from the antenna. This signal strength may be affected by many factors, including antenna shape, terrain and obstructions (which may cause complicated diffraction effects), atmospheric conditions, and, of course, distance. The effects of atmospheric conditions and distance, however, are fairly predictable. What we may need to know is, all other factors being equal, in what directions is the signal strong and in what directions is it weak?

The approach, then, is to pick a single distance (say, 25 km), and travel around a circle of this radius from the antenna, measuring the signal level for the different compass directions. The independent variable is the direction, and the dependent variable is the signal strength. The best way to represent such data is on a polar graph. A typical result is shown in figure 9-6.

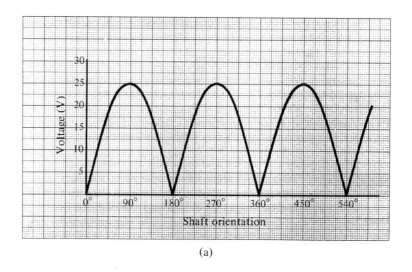

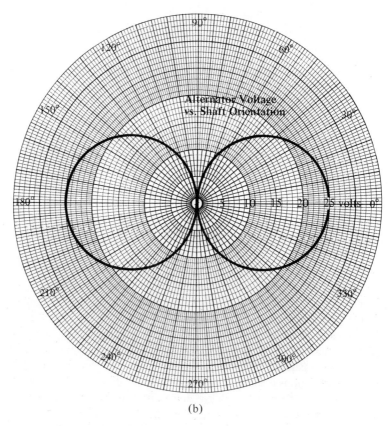

Figure 9-4. *The same data plotted on a Cartesian graph (a) and a polar graph (b).*

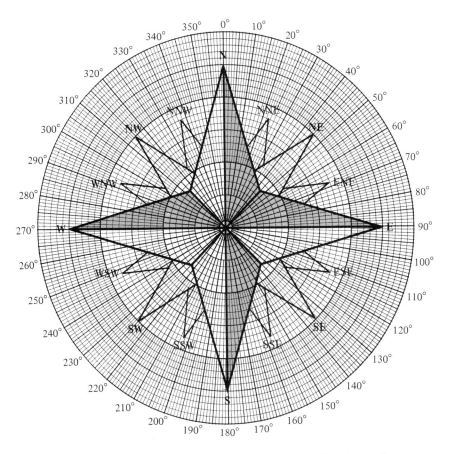

Figure 9-5. *The relationship between compass direction and angular headings.*

PPI Radar Scope

An important example of a polar graph that displays compass direction is the *plan position indicator* (PPI) radar scope. Figure 9-7 shows the basic system. A rotating antenna sends out a high-frequency radio signal that is reflected by any object in its path (bird, airplane, storm cloud, etc.). This reflected signal is picked up by the antenna, amplified, and sent to the display unit. Here an electron beam rotates around a picture tube in exact synchronization with the antenna's rotation. This beam leaves a spot on the screen each time a reflected signal has been received. The angular position of the spot from the center gives the compass direction of the object from the antenna. The radial position of the spot gives the object's distance from the antenna. This distance is calculated electronically by the instrument itself, based on the length of time between the

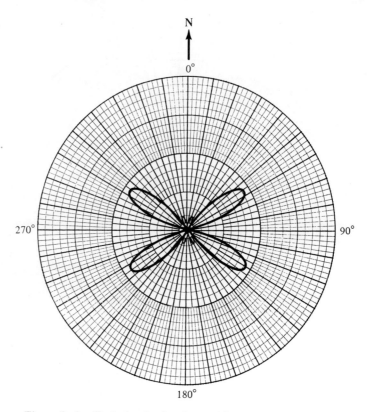

Figure 9-6. *Variation in signal strength surrounding a transmitting antenna at a 25-km distance. Signal strength is in arbitrary units.*

outgoing radar signal and the returning reflected signal. The display, then, is truly a polar graph since it gives both the direction and the distance in that direction.

Time as the Angular Coordinate on a Polar Graph

Must a polar graph always have an angle as one of the variables? No. One common procedure is to plot time as the angular coordinate, with a 24-hour day corresponding to one complete revolution on the polar graph. This gives a convenient scale, since

$$\frac{360°}{24\,\text{h}} = 15\ \text{deg/h}$$

One angular degree, then, corresponds to 1/15 h, or 4 min. Such graphs are useful for recording the variation in some quantity over an entire day. As an example, figure 9-8 shows the temperature of a certain annealing oven as it varies over a 24-h period.

Polar Graphs with Curved Spokes

Figure 9-9 shows another instance where time is used as the angular coordinate. The instrument shown is a *psychrometer,* which detects relative humidity. (Monitoring humidity is critical in the paper and textile industries.) Actually, this instrument records two temperatures—one made with a dry thermometer and one with a wet thermometer. The two readings are identical when the humidity is 100 percent. The greater the difference between the two temperature readings, the lower the humidity. In this case the pens travel radially while the paper rotates under them, just like the arm on a record player. The paper makes one complete revolution in 24 h. The spokes are 1 h apart.

Looking closely at this photograph, we see that the radial spokes on the polar paper are not straight lines. Because of the way the recording pens are mounted, they travel in arcs. The paper they write on must show these same arcs if the graph is to be read accurately. As a result, specially printed polar paper must be used with such recording instruments.

Polar graphs are also useful for simplifying certain kinds of engineering and scientific calculations. Two important cases are the calculation of electrical impedance and the design of cams. Since this book is primarily concerned with the problems of measurement and the representation of the results, we will forgo any further discussion of such applications here.

Summary

Although graphs are usually drawn according to a *Cartesian coordinate system,* a number of other systems have also been devised. One of the most important of these is the *polar coordinate system.* Polar graphs are commonly used when one of the variables is an angle—compass direction, for instance. Time is also frequently plotted as an angular variable. Polar graphs are often seen with recording instrumentation, since the mechanical recording device need be no more complicated than a record turntable.

(a)

(b)

Figure 9-7. *PPI (plan position indicator) radar system: (a) rotating antenna; (b) polar display. [Courtesy of (a) RCA, (b) Westinghouse Electric Corporation, Defense Group]*

Figure 9-8. *Time as the angular coordinate on a polar graph: variation in temperature in an annealing oven during a 24-h period.*

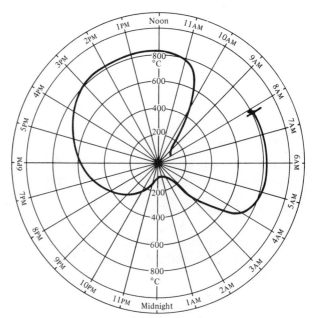

Figure 9-9. *A psychrometer assembly with a 24-h polar recorder. (Courtesy of Honeywell Inc.)*

REVIEW QUESTIONS

1. What is a Cartesian coordinate system?
2. What is a polar coordinate system?
3. What is meant when we say that two circles are *concentric*?
4. What is the range of the angular variable on standard polar coordinate graph paper?

5. Explain how data points are plotted on polar coordinate paper.
6. How are angles greater than 360° indicated on a polar graph?
7. Can a Cartesian coordinate graph be converted to a polar graph of the same data?
8. What is the standard rotation direction for increasing angular coordinate on a polar graph?
9. What special rule applies when the angular coordinate indicates compass direction?
10. What is the function of a PPI radar scope?
11. Outline the principle of operation behind a PPI radar system.
12. What quantity other than an angle is often plotted as the angular variable on a polar graph?
13. What is a psychrometer?
14. Why does polar graph paper sometimes have curved spokes rather than straight ones?

EXERCISES

1. Draw a polar graph of the following data:

r	14.9	4.0	2.0	1.3	1.1	1.0	1.1	1.3	2.0	4.0	14.9
θ	30	60	90	120	150	180	210	240	270	300	330

2. For the spiral spring discussed in example 9-1, find the torque at angles of 45°, 186°, 417°, and 890°. Is it more accurate to interpolate from the data or from the graph?

3. Draw a polar graph of the following data. Notice that the same value of r occurs for a number of different values of θ.

r	θ (deg)								
6.0	0	90	—	180	—	270	—	360
5.9	5	85	95	175	185	265	275	355
5.6	10	80	100	170	190	260	280	350
5.2	15	75	105	165	195	255	285	345
4.6	20	70	110	160	200	250	290	340
3.0	30	60	120	150	210	240	300	330
1.0	40	50	130	140	220	230	310	320
0	45	—	135	—	225	—	315	—

4. The following equation describes a polar function called a "cardiod."

$$r = 6(1 - \sin \theta)$$

Using a calculator, find the values of r for values of θ between $0°$ and $360°$ in 10-deg jumps. Use this data to graph the curve.

5. Draw a polar graph of the equation

$$r = 0.012\theta$$

where θ assumes values between $0°$ and $1\,000°$.

6. The following graph shows how an engine valve is to open and close as the engine rotates. The valve is to be driven by a rotating cam with a minimum radius of 2.00 cm. Using polar graph paper, draw the shape of the cam.

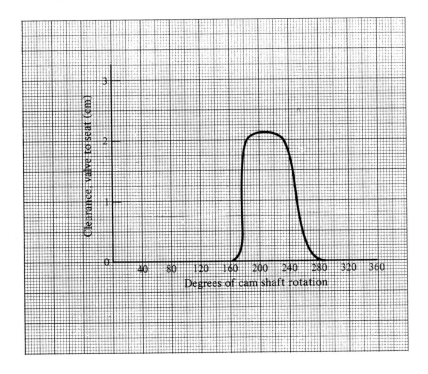

CHAPTER 10

Logarithmic Graphs

Data Spanning a Large Range

We often need to graph data that spans a very large range. It may happen that the largest data value is as much as 1 000 times, 10 thousand times, or even 1 million times the smallest value. An example of such a set of data is shown in table 10-1.

Suppose that we want to represent the table's information on a bar graph.

Table 10-1. *Half-lives of the Uranium Isotopes*

Isotope	Half-life
U = 227	1.3 min
U = 228	9.3 min
U = 229	58 min
U = 230	20.8 d
U = 231	4.2 d
U = 232	73.6 y
U = 233	1.62×10^5 y
U = 234	2.48×10^5 y
U = 235	7.13×10^8 y
U = 236	2.39×10^7 y
U = 237	6.75 d
U = 238	4.51×10^9 y
U = 239	23.5 min
U = 240	14.1 h

Our paper is 20 cm (about 8 in.) wide. Our scale has to be based on the largest data value, which is 4.51×10^9 y. This gives

$$\frac{4.51 \times 10^9 \text{ y}}{20 \text{ cm}} = 2.225 \times 10^8 \text{ y/cm}$$

We can round this off to a scale of

$$1 \text{ cm} = 2.5 \times 10^8 \text{ y}$$

Unfortunately, this choice of scale leads to problems. A data value of 162 000 years becomes 6.48×10^{-4} cm, which is certainly much too small to plot! In fact, only two of the entries in the table would show up clearly on this scale. The smallest data value, 1.3 min, would have to be represented by a bar only 9.9×10^{-15} cm long!

Logarithmic Graphs

Fortunately, we have a way to get around this difficulty. Rather than plotting the actual data, we can plot the *power of ten* of the data. When we do so, we have a *logarithmic graph*. The result, with all the units converted to seconds, is shown in figure 10-1.

DEFINITION | *A logarithmic graph is one in which the power of ten, or the logarithm ("log"), of the data is graphed rather than the actual data.*

Powers of Ten and Logarithms

Notice that the words "logarithm" and "power of ten" have the same meaning. You should already be familiar with the idea of using powers of ten to represent very large or very small numbers. The number "one million," for instance, may be written

$$1\ 000\ 000 = 10^6$$

The power of ten here is 6, which is also the logarithm of 1 000 000. Similarly, the number "one one-thousandth" may be written

$$0.001 = 10^{-3}$$

Here the power of ten is -3, which is also the logarithm of 0.001.

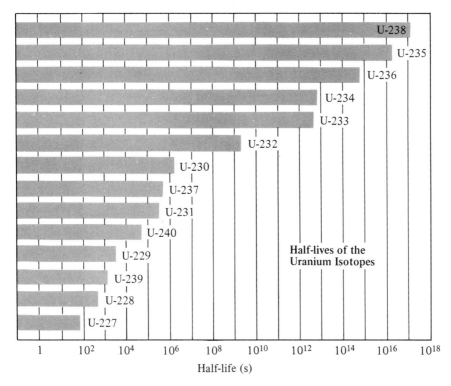

Figure 10-1. *The data of table 10-1 represented on a bar graph with a logarithmic scale.*

We may summarize the meaning of the logarithm this way: if some number N can be written as 10 raised to a power,

$$N = 10^k \tag{10-1}$$

then the logarithm of N is just the exponent on the 10:

$$\log N = k \tag{10-2}$$

Table 10-2 lists the integer powers of ten and their logarithms. Notice that we cannot take the logarithm of 0 or any negative number. If a number falls between 0 and 1 (in other words, if the number is a fraction), its logarithm is negative. If a number is larger than 1, its logarithm is positive.

In graphing data, we will need to find logarithms of numbers other than exact powers of ten. Calculating such logarithms from scratch is fairly involved, so we will not go into the details here. Fortunately, most scientific calculators do the same thing with the push of a single key. If you have such a calculator, read the manual carefully and acquaint yourself with the use of the logarithm function. One note of caution: many calculators have *two* logarithm functions. One is the base-ten logarithm we are using here, and the other is the base-*e*

Table 10-2. *The Integer Powers of Ten and Their Logarithms*

Number	Power of Ten	Logarithm
1 000 000 000	10^9	9
100 000 000	10^8	8
10 000 000	10^7	7
1 000 000	10^6	6
100 000	10^5	5
10 000	10^4	4
1 000	10^3	3
100	10^2	2
10	10^1	1
1	10^0	0
0.1	10^{-1}	-1
0.01	10^{-2}	-2
0.001	10^{-3}	-3
0.000 1	10^{-4}	-4
0.000 01	10^{-5}	-5
0.000 001	10^{-6}	-6
.	.	.
.	.	.
.	.	.
0	$10^{-\infty}$	$-\infty$

logarithm that finds frequent use in engineering calculations. While either one, if used consistently, can be used for drawing logarithmic graphs, you can avoid a great deal of confusion by using only the base-ten logarithm for this purpose.

Using a Calculator to Find Logarithms

▶ **Example 10-1: Logarithms of some numbers**
Using a scientific calculator, find the base-ten logarithms of the following numbers:

3.00	9.37×10^5
6.00	4.11×10^{-3}
3 00$\overline{0}$	10$\overline{0}$
6 00$\overline{0}$	0

On many calculators, you simply key in the number and press the LOG key. You should get the following results:

$$\log 3.00 = 0.477$$

$$\log 6.00 = 0.778$$

Notice that the logarithm of 6.00 is *not* double the logarithm of 3.00.

$$\log 3\,00\bar{0} = 3.477$$

$$\log 6\,00\bar{0} = 3.778$$

These two results are exactly 3 larger than the previous two results. No accident, this comes about because $3\,00\bar{0}$ is 1 000 times as large as 3.00, and the logarithm of 1 000 is 3.

$$\log 9.371 \times 10^5 = 5.972$$

$$\log 4.11 \times 10^{-3} = -2.39$$

$$\log 10\bar{0} = 2.00$$

$$\log 0 = \text{error}$$

Most calculators give some type of error indication on this last calculation. There is no such thing as the logarithm of 0.

Notice that the original numbers, excluding 0, went from a low of 4.11×10^{-3} to a high of 6 000. If these had been measurement data, it would be impossible to graph them in the conventional way. The logarithms, however, vary from a low of -2.39 to a high of 5.972, a range that can be easily plotted on a graph. Taking logarithms of a set of data spreads out the small values and squeezes the large values closer together. ◄

Although the calculator is a great convenience in finding logarithms, it is not an absolute necessity. Logarithms can also be found from tables. Table 10-3 is a listing of the logarithms of numbers between 1 and 10. We see from this table, for instance, that the logarithm of 7.3 is 0.863.

What if the number we are looking for is larger than 10? No problem. We simply write the number in power-of-ten notation, look up the logarithm of the resulting number between 1 and 10, then add this to the power of ten. The next example shows how.

Table 10-3. *Base-ten Logarithms of Numbers 1 through 9*

	0.0	0.1	0.2	0.3	0.4	0.5	0.6	0.7	0.8	0.9
1	0.000	0.041	0.079	0.114	0.146	0.176	0.204	0.230	0.255	0.279
2	0.301	0.322	0.342	0.362	0.380	0.398	0.415	0.431	0.447	0.462
3	0.477	0.491	0.505	0.519	0.532	0.544	0.556	0.568	0.580	0.591
4	0.602	0.613	0.623	0.634	0.644	0.653	0.663	0.672	0.681	0.690
5	0.699	0.708	0.716	0.724	0.732	0.740	0.748	0.756	0.763	0.771
6	0.778	0.785	0.792	0.799	0.806	0.813	0.820	0.826	0.833	0.839
7	0.845	0.851	0.857	0.863	0.869	0.875	0.881	0.887	0.892	0.898
8	0.903	0.909	0.914	0.919	0.924	0.929	0.935	0.940	0.945	0.949
9	0.954	0.959	0.964	0.969	0.973	0.978	0.982	0.987	0.991	0.996

Using Logarithm Tables

▶ **Example 10-2: Logarithms of some more numbers**
Using table 10-3, find the logarithms of the following numbers:

$$3\ 700$$

$$0.000\ 12$$

The first number may be written in power-of-ten notation as follows:

$$3\ 700 = 3.7 \times 10^3$$

Looking up the logarithm of 3.7, we get

$$\log 3.7 = 0.568$$

Adding on the 3 (the power of ten), our result is

$$\log 3\ 700 = 3.568$$

The other number is handled similarly, except that algebraic signs are involved. In power-of-ten notation, the number becomes

$$0.000\ 12 = 1.2 \times 10^{-4}$$

Then

$$\log 1.2 = 0.079$$

Adding −4, we get

$$\log 0.000\ 12 = -3.921 \blacktriangleleft$$

Now that we have discussed how to find logarithms, and what effect they have on a set of data, let's look at the process of drawing a logarithmic functional graph.

Drawing a Logarithmic Functional Graph

▶ **Example 10-3: Fluid flow rate**
There are many methods for measuring fluid flow rate. Figure 10-2 shows one of them, an *orifice manometer*. A plate with a hole (the orifice) is placed in the pipe, and this causes a pressure drop as the fluid flows through the constriction. A mercury manometer measures the pressure difference between the upstream and downstream sides of the orifice plate. This pressure difference is an indication of the flow rate.

As with any measuring instrument, the device must be calibrated for use. This can be done by introducing known flow rates in the pipe, then recording the corresponding manometer readings. A graph of this set of data becomes the calibration curve for the instrument. In other words, such a graph can be used to convert any subsequent manometer reading to the corresponding flow rate.

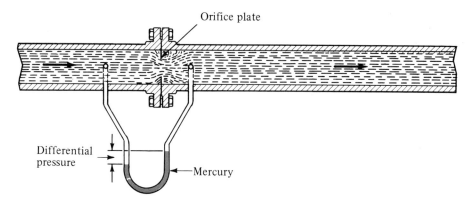

Figure 10-2. *An orifice manometer for measuring fluid flow rate.*

Table 10-4 shows some typical results. The flow rate has a range of about one power of ten, from 0.72 to 7.43 l/s. There is no problem in plotting this variable in the usual way. The pressure measurement, on the other hand, varies over a range of two powers of ten, from 1.4 to 148. Moreover, the smaller values are crowded together while the large ones are spread apart: the first four values go only as high as 16.2, while the next four go up to 148. These characteristics make the set of data a good candidate for a logarithmic graph.

The first step is to find the logarithm of each of the pressure measurements. This, of course, can be done by using either a calculator or table 10-2. The results are listed in table 10-5, which will be our working table for the graph.

The pressure here is the dependent variable, since a change in flow rate will result in a change in pressure. Ordinarily, we would then plot the pressure on the vertical axis and the flow rate on the horizontal axis. If, however, we plan to use the graph as a calibration curve, we will probably want to reverse the two. This is because the *user* will first read the pressure

Table 10-4. *Measurements Taken with an Orifice Manometer*

Flow Rate (l/s)	Orifice Manometer Reading (mm Hg)
0.72	1.4
1.08	3.1
1.80	8.7
2.46	16.2
4.04	43.7
5.31	75.5
7.15	137
7.43	148

Table 10-5. *Working Table for Graphing the Flow Rate versus the Logarithm of the Pressure in Table 10-4*

Flow Rate (l/s)	Pressure (mm Hg)	Logarithm of Pressure
0.72	1.4	0.146
1.08	3.1	0.491
1.80	8.7	0.940
2.46	16.2	1.21
4.04	43.7	1.64
5.31	75.5	1.88
7.15	137	2.14
7.43	148	2.17

from the instrument, then go to the curve to find the corresponding flow rate. In practice, the user's result for the flow rate will depend on the pressure measurement she or he has made. We will therefore plot the flow rate here as the dependent variable and the logarithm of the pressure as the independent variable. The result is shown in figure 10-3. ◄

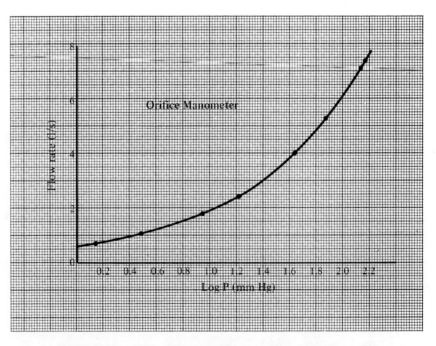

Figure 10-3. *Calibration curve for an orifice manometer. This graph is based on the data in tables 10-4 and 10-5.*

Graphing with Logarithmic Graph Paper

Figure 10-3 has a drawback in that it requires a table of logarithms or a calculator to use it. If we read the pressure differential from the manometer, we have to take the *logarithm* of this number before we can read the flow rate from this graph. Although finding logarithms is not particularly difficult, sometimes it can be inconvenient. Is there some way to draw a logarithmic graph but avoid this inconvenience in its use?

Yes. If we sketch the graph on *logarithmic graph paper,* we can read it directly without using a logarithm table. Using log paper also simplifies the procedure for drawing the graph. Logarithmic graph paper, or log paper, is characterized by a grid of lines whose spacing varies according to the logarithm of the number represented. Since logarithms spread out the smaller data values and compress the larger ones closer together, the spacing of the grid lines follows this same pattern. The paper may have such a logarithmic scale on just one axis, or on both axes. If the log scale is on only one axis, the paper is called *semilog* paper. If the log scale is on both axes, the paper is *log-log* paper.

Besides the semilog or log-log designation, the paper is also described by the number of log cycles on each axis. For instance, semilog paper is available in single-cycle, two-cycle, three-cycle, or four-cycle form. Log-log paper comes in one-cycle-by-one-cycle, one-cycle-by-two-cycles, two-cycles-by-three-cycles, and other combinations.

Log paper forces the scale on the user. Numbers are already printed on the paper, and the 1 can represent only 1 or 10 or 100 or some other integer power of ten—but it can never represent 2, for instance. Similarly, the 4 on the paper can be used only as 0.4, 40, 4 000, or some other integer power-of-ten multiple of 4. Thus the paper must be chosen to fit the data. If the data spans a range of less than a factor of 10, you can usually use single-cycle log paper. If the range is greater than a factor of 10 but less than a factor of 100, you will need two-cycle paper, and so on. The next example shows how to choose the correct paper and draw a logarithmic graph on it.

Graphing with Log Paper

▶ Example 10-4: Radioactive decay of iodine-130

Table 10-6 lists data on the rate of emission of beta particles from a small sample of the radioisotope iodine-130. This particular isotope can be created in the laboratory (although not too easily), and it is found to be radioactive. It transforms itself into the isotope xenon-130 through a nuclear process known as negative beta decay. Xenon-130 is a stable isotope and in fact can be found in trace amounts in the atmosphere.

The rate at which iodine-130 decays into xenon-130 can be measured

Table 10-6. *Variation of the Rate of Beta-Particle Detection from a Sample of Iodine-130*

Counts/min	Elapsed Time (h)
896	0.0
538	9.2
353	16.8
237	24.0
110	37.9
86	42.3
46	53.7
23	65.9

by using a particle detector to intercept the beta particles, and an electronic counter to count them. Table 10-6 shows the counting rate, in particles per minute, as time goes on and less and less of the original sample remains. The problem here is to graph this data on log paper.

We see that the elapsed time varies from 0.0 to 65.9 h. Since there is no way to plot the logarithm of 0, we will put this variable on a conventional (nonlogarithmic) scale.

The other variable, the counting rate, varies from 23 to 896 counts/min. Plotting this quantity will require two log cycles, which will span the ranges 10 to 100 and 100 to 1 000. Notice that there is no way to put all of this data onto a one-cycle logarithmic graph.

The elapsed time is the independent variable, while the counting rate is the dependent variable. This means that the logarithmic scale will run vertically. When we label the axes and plot the data on two-cycle semilog paper (or three-cycle paper with one cycle removed), we get the result shown in figure 10-4.

Examine this graph carefully to see how the data was plotted. It is particularly important to notice that the logarithmic scale is a nonlinear scale. In other words, the distance between 10 and 20 is larger than the distance between 20 and 30, which in turn is much larger than the distance between 220 and 230. Although the numbers themselves are plotted directly, any number's *position* on the log paper is based on the logarithm of that number, and so the larger values are squeezed closer together. Between 10 and 20, we see 10 divisions; each division then represents 1 count/min. This is true until we get to 40, where the scale changes. Between 40 and 50 we see only 5 divisions, each of which represents 2 counts/min. The scale again changes when we reach 100. Between 100 and 400, each division represents 10 counts/min. After 400, each division represents 20 counts/min. Because of this variable scale, it is very easy to make a mistake in plotting a logarithmic graph, and each point should be checked several times before the interpolated curve is drawn. ◄

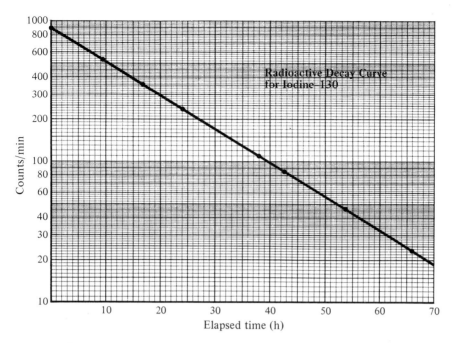

Figure 10-4. *A graph of the data in table 10-6 on two-cycle semilog paper.*

Semilog Graphs

Figure 10-5 shows a logarithmic graph drawn on three-cycle semilog paper. Here the log scale is used for the independent variable. Notice that although this variable assumes rather large values, the beginning of each log cycle still represents an integer power of ten.

Figure 10-6 shows a six-cycle semilog graph. A graph with this many log cycles may require actually drawing the grid, or taping together several pieces of the available semilog paper. Notice that the largest data value in the figure is nearly a million times larger than the smallest value. With data that spans this great a range, it is totally impossible to draw a conventional graph.

Log-log Graphs

In some situations, we might want to plot both variables on a logarithmic scale. For example, the data in table 10-7 describes the operation of an electronic solid-state device known as a *varistor*. We see that a small increase in voltage on

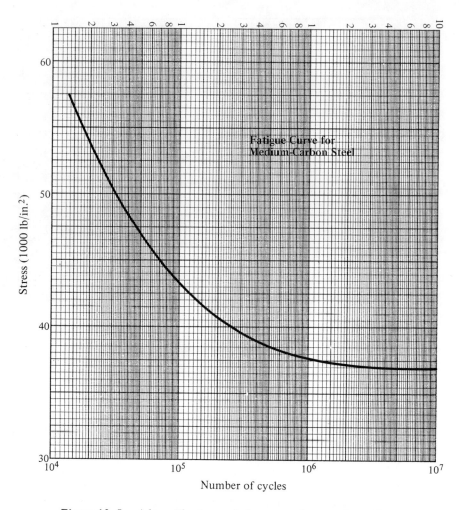

Figure 10-5. *A logarithmic graph drawn on three-cycle semilog paper.*

the varistor produces a fairly large increase in current. The current, then, is the dependent variable. A graph of the data, on log-log paper, is shown in figure 10-7. Notice that the voltage requires two log cycles while the current requires five.

Log-log Graphs and Empirical Equations

One other use for log-log graphs is to find empirical equations. We have already seen, in chapter 9, that an empirical equation can be found by making an edu-

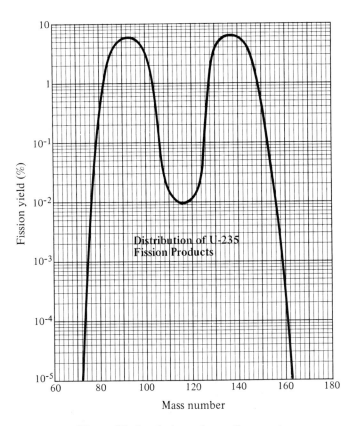

Figure 10-6. *A six-cycle semilog graph.*

Table 10-7. *Voltage-Current Data for a Varistor*

Voltage (V)	Current (A)
10.0	1.3×10^{-4}
16.2	6.5×10^{-4}
25.0	2.8×10^{-3}
46.7	2.4×10^{-2}
62.1	6.3×10^{-2}
82.9	0.17
100	0.33
122	0.71
187	3.5
230	9.2

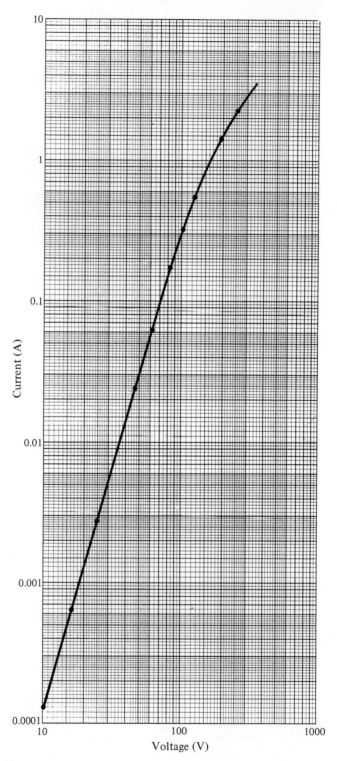

cated guess at the form of the relationship, then seeing if this guess gives a straight-line graph. Often the procedure can be streamlined by using a log-log graph.

It turns out that a set of data graphed on log-log paper will yield a straight line if the data satisfies any equation of the form

$$y = mx^k \qquad (10\text{-}3)$$

where m and k are numerical constants. For example, all the following equations give straight lines when graphed on log-log paper:

$y = 3x^2$ ($m = 3, k = 2$) $\qquad$ $y = 1.9x$ ($m = 1.9, k = 1$)

$y = -7.21x^3$ ($m = -7.21, k = 3$) $\qquad$ $y = \dfrac{4.96}{x}$ ($m = 4.96, k = -1$)

$y = x^{1.32}$ ($m = 1, k = 1.32$) $\qquad$ $y = -x^{-2}$ ($m = -1, k = -2$)

Through one graph, then, we can test a great many potential guesses.

Suppose that we graph a set of data on log-log paper and find that it does yield a straight line. This tells us that the empirical equation is

$$y = mx^k$$

The problem is then to find m and k. To get k (the exponent), we can pick two points on the line, then do the following calculation:

$$k = \frac{\log y_1 - \log y_2}{\log x_1 - \log x_2} \qquad (10\text{-}4)$$

This calculation tells us the correct *form* of the empirical equation. We may then go back to the techniques of chapter 8 to find m. Let's look at an example.

Finding the Empirical Equation from a Log-log Graph

▶ **Example 10-5: Some nonlinear data**
The problem is to find the empirical equation describing the following set of measurement data:

y	x
78.6	1.12
13.2	2.73
4.14	4.88
0.593	12.9
0.202	22.1

◀ Figure 10-7. *Current-voltage characteristics of a varistor on a two-by-five cycle log-log graph, plotted from the data in table 10-7.*

A graph of this set of data is shown in figure 10-8. Notice that these points fall very nicely on a straight line. The equation is therefore of the form $y = mx^k$. To find the exponent k, we can use any two points on this line. Remember that the coordinates of these points are the *logarithms* of the data values. Figure 10-8 indicates the second and fourth data points. Then, from equation (10-4),

$$k = \frac{\log 13.2 - \log 0.593}{\log 2.73 - \log 12.9}$$

$$= \frac{1.121 - (-0.227)}{0.436 - 1.111}$$

$$= \frac{1.348}{-0.675}$$

$$k = -2.00$$

This gives us an empirical equation of

$$y = mx^{-2}$$

or, equivalently,

$$y = \frac{m}{x^2}$$

There are a number of ways to find m. We can calculate the values of $1/x^2$ from the original data, draw a conventional (nonlogarithmic) graph of y versus $1/x^2$, then read m as the slope of this line. We can find the center of gravity of y, the center of gravity of the calculated values of $1/x^2$, then use equation (8-8). Or, more directly, we can observe that

$$m = yx^2$$

for any original data points. For the first point, this gives

$$m = (78.6)(1.12)^2 = 98.60$$

For the second point, it is

$$m = (13.2)(2.73)^2 = 98.38$$

and for the third point, it is

$$m = (4.14)(4.88)^2 = 98.59$$

The other two values are 98.68 and 98.66. A reasonable estimate for m, then, is 98.6. The empirical equation is

$$y = 98.6x^{-2}$$

You can verify that this equation will generate the original values of y from the values of x in the data table. ◀

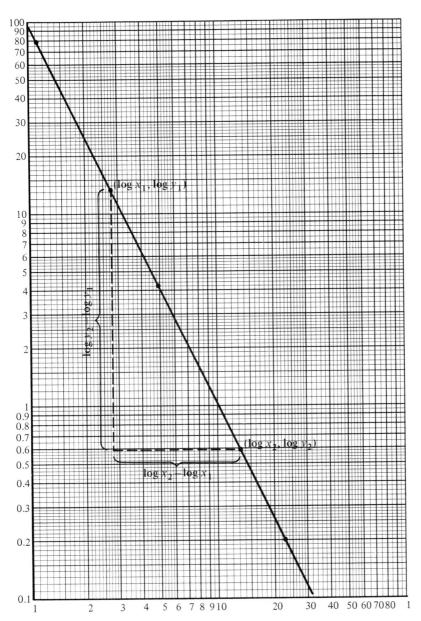

Figure 10-8. Use of a log-log graph to find the empirical equation for a set of data (see example 10-5).

Summary

A set of measurement data often spans a greater range than can be represented conveniently on a conventional graph. In such cases, it is best to plot the *logarithm* of the troublesome variable rather than the variable itself. This can be done in two ways: either by calculating logarithms from the data and plotting these values, or by using logarithmic graph paper.

Logarithmic graph paper comes in two basic forms: *semilog* and *log-log*. The paper is further characterized by the number of logarithmic cycles on each axis. Since logarithmic paper forces the scale on the user, the paper must be chosen according to the range in the data being graphed.

Log-log paper may also be used to streamline the procedure for finding an empirical equation. If a log-log graph is a straight line, the original data is a power function. The exponent on the independent variable is then the slope of the log-log graph. One graph therefore eliminates the need for repeated guesses and trials of the form of the empirical equation. In fact, there is generally no practical alternative to the log-log graphical procedure if the independent variable has a fractional exponent.

REVIEW QUESTIONS

1. How is a logarithmic graph different from a conventional graph?
2. Why are logarithmic graphs used?
3. What is the logarithm of 10?   100?   100 000?   0.01?
4. How are logarithms of numbers other than integer powers of ten found?
5. What is the logarithm of 2.7?   4.9?   9.0?
6. What is an orifice manometer used for?
7. Why are the dependent and independent variables usually interchanged when a calibration curve is drawn?
8. Can a logarithmic graph be drawn without logarithmic graph paper?
9. It was pointed out that logarithmic graph paper forces the scale on the user. What does this mean?
10. What is the difference between semilog paper and log-log paper?
11. Why is it easier to make a mistake in drawing logarithmic graphs than in drawing conventional graphs?
12. If a log-log graph is a straight line, what does this say about the empirical equation of the data?

13. Why might a log-log graph not be a straight line?
14. Outline the procedure for finding an empirical equation from a log-log graph of the data.

EXERCISES

1. Find the logarithms of the following numbers:
   a. 6.7
   b. 670
   c. 0.006 7
   d. 2.3
   e. 8720
   f. 100 000 000
   g. 2.7×10^6
   h. 5.9×10^{-8}
   i. 5.9×10^8
   j. 1.000

   [Answers: (h) −7.23; (i) 8.77]

2. Decide how many log cycles will be needed to graph data that span the following ranges.
   a. 1.71 to 98.2
   b. 0.161 to 76.3
   c. 263 to 88 100
   d. 19.2 to 137
   e. 0.000 31 to 0.046
   f. 5.3×10^{-2} to 43
   g. 6.7×10^3 to 7.8×10^4
   h. 1.76×10^{-5} to 8.37×10^{-5}

   [Answers: (a) two; (d) two; (f) four]

3. Draw a semilog graph of the following data (a) by calculating the logarithms of the x values, and (b) by using semilog paper.

y	x
2.10	10.6
6.70	67.7
10.5	196
18.7	1 140
26.3	4 230

4. Graph the data in table 10-5 on semilog paper.

5. Consult an almanac to find data on the U.S. population at each census between the years 1790 and 1970. Graph the data on a semilog graph.

6. Draw a semilog graph showing the increase in world population between 1650 and today. Consult an almanac or encyclopedia to get the data.

7. Graph the data in table 10-2 (a) on semilog paper, and (b) on log-log paper. Based on the log-log graph, write the correct form for the empirical equation.

8. Draw log-log graphs for the data in exercise 5, chapter 8, parts b, e, f, and i. Use these graphs to find the empirical equations.

CHAPTER 11

Indicators

A vast number of measuring devices are in common use, and new ones are being patented each year. It would not be practical to try to discuss all such devices in a book like this. What we can do, however, is look at some of the features that many of these instruments have in common and to mention some of the possible pitfalls in their use. In this chapter, we will examine the standard methods for numerical indication in instruments, and in chapter 12 we will look at some of the common sensing techniques.

Sensors and Indicators

We have already seen that a measurement is the comparison of a physical quantity with a standard. Most instruments make this comparison in two steps: First, they *sense* the quantity being measured, and then they *indicate* the size of the quantity on some sort of scale or display. This is a useful distinction to make, since in many instruments the sensing unit is completely separate from the indicating, or display, unit. The fuel gauge on the dashboard of a car, for instance, indicates the fuel level in the tank at the rear of the vehicle. The sensing unit is in the tank, while the indicator is on the dashboard. The two are quite separate, yet together they form one instrument. A more dramatic example, shown in figure 11-1, is the control room for the University of California's Bevalac, which is a type of particle accelerator used for experiments in nuclear physics. The panel contains a large number of indicator and display units whose sensors are located quite a distance away.

DEFINITION | A <u>sensor</u> is a device that responds to changes in a specific physical quantity, and that has been designed to transmit information on this change to an indicator, or display unit.

Figure 11-1. *Control room at the University of California's Bevalac. (Courtesy of Lawrence Berkeley Laboratory)*

DEFINITION | An <u>indicator</u>, or <u>display</u> unit, is a device that converts information from the sensor to a numerical value that the user can read. Display units are seen as digital readouts, chart recorders, cathode ray tubes, micrometer and vernier scales, and other forms.

Micrometer Indicators

Let's begin with the micrometer indicators. The familiar micrometer caliper shown in figure 11-2 is an instrument capable of measuring dimensions as small as 0.01 mm, which, of course, is much finer than can be read by eyeballing a conventional rule. The instrument achieves this great sensitivity through a screw turning in a threaded sleeve. One complete turn of the thimble advances the screw a given amount—typically, 0.5 mm. Multiples of this distance are indicated on a fixed scale on the sleeve (or barrel) of the instrument, and are uncovered as the thimble moves. The thimble's circumference is further divided into 50 divisions, so a rotation through one of these divisions corresponds to an advance of 1/50 of 0.5 mm, or 0.01 mm. To read a micrometer, we first read the fixed scale to the last 0.5-mm division, then we read the rotating scale to get the number of hundredths of a millimetre beyond the last multiple of 0.5 mm.

We often think of the micrometer as a complete instrument when in fact it is only the indicator. But a micrometer *caliper* is a complete instrument, because it has a sensor (the anvil and spindle) as well as an indicator (the micrometer scale). Micrometer indicators are used on a variety of instruments, some examples of which are shown in figure 11-3.

Vernier Indicators

Another important type of indicator is the *vernier*. Like the micrometer, it is found in precision dimensioning instruments as well as in instruments with other applications. A vernier caliper is shown in figure 11-4. The indicator portion of the caliper consists of a rule that itself can typically be read to 0.5 mm (figure 11-5). Adjacent to this rule is a sliding scale that extends the precision to 0.02 mm. Twenty-five divisions on the sliding scale correspond to 24 divisions on the fixed scale above it. In that way, additional precision is achieved because one division on the sliding scale comes closer than the others to lining up with a division on the fixed scale. This gives a multiple of 0.02 mm, which is then added to the basic reading on the fixed scale. In figure 11-5, for instance, the reading is 41.68 mm.

A vernier indicator is never particularly easy to use, and most people need quite a bit of practice to get very good at it. Yet if extreme precision is required, a vernier instrument can give it.

Dial-type Indicators

One of the oldest and most basic display units is the conventional clock dial, which has remained unchanged for over seven centuries. The same design is seen

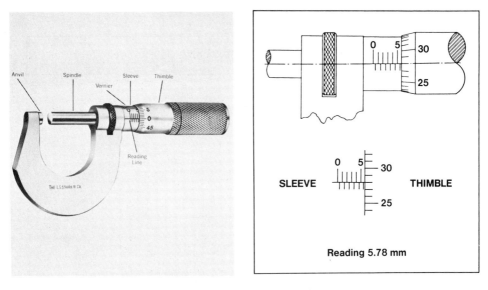

Figure 11-2. *A metric micrometer caliper. (Courtesy of The L. S. Starrett Company)*

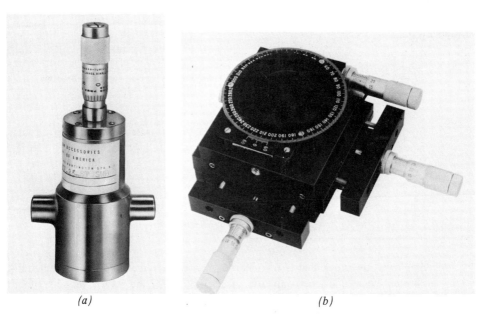

Figure 11-3. *Two instruments that use micrometer indicators: (a) precision gas-metering valve and (b) precision rotator/translator for optical measurements. (Courtesy of (a) Vacuum Accessories Corp. of America, and (b) Oriel Corporation of America)*

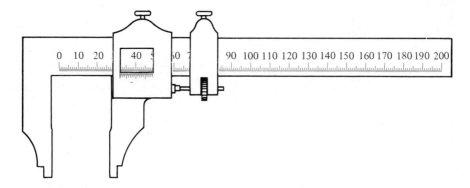

Figure 11-4. *A vernier caliper.*

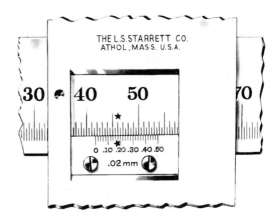

Figure 11-5. *The indicator portion of a metric vernier caliper. (Courtesy of The L. S. Starrett Company)*

in the mechanical aircraft altimeter: two hands sweeping around a circular, numbered dial with the small hand indicating altitude in thousands of feet and the large hand indicating hundreds of feet.

The d'Arsonval Movement

Similar to the clock dial is the electrical meter movement with a single hand or pointer. We will discuss only the most common of the several versions of this device; the *moving-coil,* or *d'Arsonval, movement.* This meter movement, shown in figure 11-6, is used for measuring small electrical currents, although it can also measure other quantities through an appropriate separate sensor. The current is introduced by completing a circuit through connections A and B, so the tiny moving coil becomes a part of the external circuit. This coil is wound on

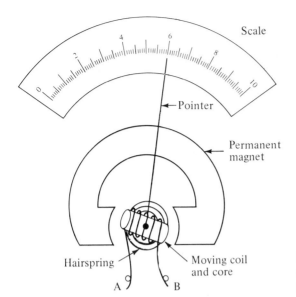

Figure 11-6. *The moving-coil, or d'Arsonval, meter movement.*

an iron core, which is supported on bearings but held in position by a hairspring (the same type of spring as used on the balance wheel of a clock or watch). Under the current's influence, the iron core becomes a magnet and rotates in an attempt to align its poles opposite to those of the permanent magnet. It is restrained in this movement by the hairspring, which provides a countertorque. Thus the coil can rotate only a certain amount before its magnetic torque is balanced by the hairspring's countertorque. This rotation is indicated on a scale by the position of a pointer. The larger the current, the greater the magnetic torque and the farther the pointer moves to the right. The zero point can be adjusted, with no current flowing, by turning a screw that alters the tension on the hairspring. It should be obvious, from the diagram, that the total movement of the pointer must be less than 180 angular degrees.

Meter Damping

Now if we were to use the meter exactly as it is shown in figure 11-6, we would soon begin to lose patience with it. Let's say we introduced a steady current of 1 mA. The pointer would swing to the right, *past* the 1 mA division, slow down, stop, swing back beyond the division, and so on, oscillating back and forth for a very long time before it finally would come to rest. Mechanical oscillations like this are characteristic of any system that balances a force against a counterforce, or a torque against a countertorque. You may have experienced this effect if you have ever driven a car with bad shock absorbers on a bumpy road. Just as a car needs good shock absorbers to prevent it from oscillating wildly on its suspension, so the meter movement needs something to prevent the moving coil and pointer from oscillating on the hairspring. This is usually achieved by including a set of aluminum plates or winding the coil on an aluminum frame where induced eddy currents counteract any abrupt motion of the coil. The process is called *damping*.

DEFINITION | *Damping* is the suppression of oscillation.

Lag Time and Time Constant

Now, the benefits of damping are not without drawbacks. The greater the damping, the slower the pointer response to a change in current. In other words, in suppressing oscillation we are forced to introduce *lag time* into the indicator.

DEFINITION | *Lag time* is the time it takes an indicator to accurately indicate a change in the quantity being measured.

Figure 11-7 is a graph of how a certain current changes in time. On this same graph, we see how the meter movement responds to this current change. The length of time it takes the meter to indicate the new current value is the lag time. If the meter damping is increased, the lag time also increases.

This same property of indicators is sometimes described by two other terms: *risetime* and *time constant*. The word *risetime* is common in electronics applications where the response of the instrument must be very fast, say, a few

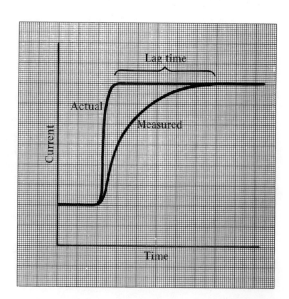

Figure 11-7. *Lag time in the measurement of current. The meter does not respond as quickly as the actual change in the current being measured.*

billionths of a second. Risetime means the same thing as lag time, as we have defined it here. The term *time constant,* however, is slightly different:

DEFINITION | The time constant of an instrument is the length of time needed for the instrument to register 63.2 percent of the actual change in the value of a measured variable.

Obviously, an instrument's time constant is related to its lag time. In fact, *time constant* is actually a better term to use, since it has a quantitative definition. As a rule of thumb, however, we can say that the lag time is approximately five time constants:

$$\text{lag time} \approx 5 \text{ time constants} \qquad (11\text{-}1)$$

This relationship is valid if the measured variable undergoes an *abrupt* change. If the change is gradual, the lag time may be quite different. The time constant, however, is a characteristic of the instrument itself and does not change very much with the conditions of its use.

Lag Time in a Current Measurement

▶ **Example 11-1: Lag time of a d'Arsonval movement**
A certain electrical meter movement has a time constant of 1.25 s. The meter is properly calibrated, then switched into a line carrying a 2.50-A current. Sketch a graph showing the meter's reading versus time.

When the meter is first connected to the current, it reads 0 A. One time constant (1.25 s) later, the reading has climbed to 63.2 percent of 2.50 A, or 1.58 A. This leaves a 0.92-A difference between the reading of the meter and the actual current. During the next 1.25 s (i.e., one more time constant), the meter reading increases by 63.2 percent of this 0.92 A, or 0.58 A. This gives us a reading, after 2.50 s, of 1.58 A + 0.58 A, or 2.16 A. The discrepancy is now 2.50 A − 2.16 A, or 0.34 A. During the next 1.25 s, the meter reading again increases by 63.2 percent of this difference, or 0.21 A. Now 3.75 s have elapsed and the reading is 2.16 A + 0.21 A, or 2.37 A. This reasoning can be repeated to get the readings at the next two multiples of 1.25 s. The results are as follows:

Time (s)	Meter Reading (A)
0	0
1.25	1.58
2.50	2.16
3.75	2.37
5.00	2.45
6.25	2.48

Notice that after 5 time constants, the meter still theoretically reads slightly below the actual current value. In practice, however, we can usually assume that this reading is close enough to be considered accurate. Nevertheless, if extreme accuracy is required, we are justified in waiting 10 time constants for the meter to reach equilibrium. The graph is shown in figure 11-8. ◄

Applications of the d'Arsonval Movement

The moving-coil meter movement is really a complete instrument having both a sensor and an indicator. The sensor is the hairspring, which determines the countertorque needed to balance the magnetic torque produced by the current. The display is a simple mechanical pointer on a scale. The movement is typically capable of handling currents of 250 microamperes (μA). To measure larger

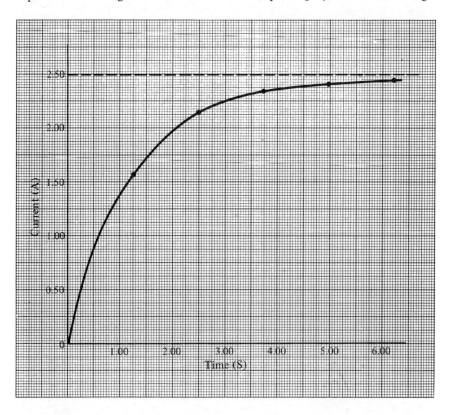

Figure 11-8. *Response of a meter with a time constant of 1.25 s. The lag time is approximately 6 s.*

currents, the movement is connected to a current divider; for smaller currents, it is connected to an amplifier. In these cases the current divider or amplifier acts as the sensor, and the entire meter movement can be considered to be the indicator.

Many other instruments use the moving-coil movement to display measurements other than current and voltage. A pH meter, for instance, gives a measurement of the acidity or alkalinity of a liquid. An automotive engine analyzer can measure dwell angle, engine revolutions per minute, and resistance of components such as the coil, points, and wiring. In fact, any quantity that can be converted to an electric current can be measured with a moving-coil meter, and in such cases the movement becomes the indicator.

Scale Linearity

A moving-coil meter's scale may be *linear* or *nonlinear*.

DEFINITION | *A linear meter scale is one in which the size of the scale divisions stays the same from one end of the scale to the other.*

DEFINITION | *A nonlinear meter scale is one in which the scale divisions get progressively smaller from one end of the scale to the other.*

Figure 11-9 shows examples of these two types of scales. The linear meter scale is obviously the simpler one to read. A linear scale is used any time the sensor responds to the measured variable in a linear way—that is, in a way that can be described by a linear equation. Often, however, a sensor responds in a more complicated way, a situation that requires a more complicated scale. Ohmmeters and ac ammeters and wattmeters are common instruments that use nonlinear scales. The point here is that the scale must be designed to reflect what the sensor does; if the sensor has a nonlinear response, the scale also must be nonlinear. Because nonlinear scales are more difficult to read than linear scales, we should take extra time and care when we use them. The situation is very similar to the problem of reading a logarithmic graph.

Accuracy and Nonlinear Scales

▶ **Example 11-2: Reading an ohmmeter**
The face of an ohmmeter is shown as it indicates two different resistances. The resistances are in units of ohms (Ω).

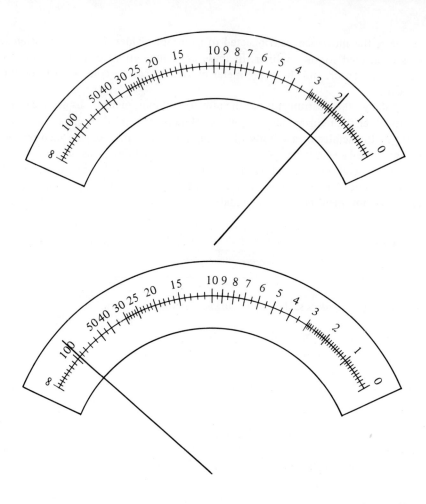

Because the scale is nonlinear, the ohmmeter is more sensitive for small resistances than for large ones. In the upper diagram, the meter reading is 1.80 ± 0.05 Ω. In the lower diagram, the pointer is in a region where the scale divisions are much closer together. In fact, the divisions come so close together that some have to be omitted for the sake of clarity. The reading here is 95 ± 3 Ω. Notice that the reading's accuracy depends on which end of the scale it falls in. If this ohmmeter has several ranges that can be selected by a switch, we should obviously use a range that puts the pointer toward the right-hand side of the scale. ◄

Parallax

Regardless of whether the scale is linear or nonlinear, meter readings are susceptible to an error known as *parallax*. Parallax errors come about because the

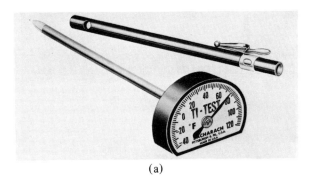

(a)

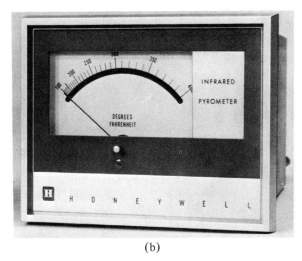

(b)

Figure 11-9. *Meter scales may be linear (a) or nonlinear (b). (Courtesy of (a) Bacharach Instrument Company, and (b) Honeywell Inc.)*

pointer does not move directly on the scale but rather at some distance in front of it. If the pointer is not viewed from a direction exactly perpendicular to the scale, the reading will appear to be higher or lower than it really is.

DEFINITION | *Parallax error is the error caused by viewing a meter pointer at other than a right angle to the scale.*

Figure 11-10 shows a top view of a meter's pointer and scale, and how their separation gives rise to parallax error. You may have noticed this problem if you have ever tried to read a car's speedometer from the passenger seat, or a public tower clock from a point on the ground below it.

Manufacturers of high-quality instruments use two different devices to

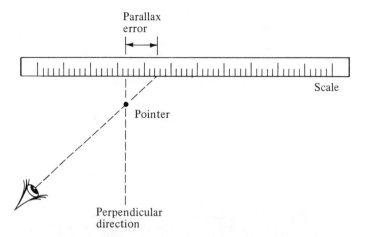

Figure 11-10. *Parallax error arises because the pointer of a meter movement is not on the scale, but in front of it. Care must be taken to read the meter from a direction perpendicular to the scale face.*

help the user keep parallax error to a minimum. One of these devices is the *mirrored scale*. While it is difficult to guess whether you might be reading this book at a 90° angle to your line of sight, it is simple to tell if you are looking into a mirror at a 90° angle. How? Simply by seeing an image of your eye. Moreover, if you hold a pencil upright in front of a mirror, the pencil will hide its own reflection when you view it from a 90° angle to the mirror. If you have a mirror on a meter's scale, you can be sure you are reading from a 90° angle when the pointer hides its own reflection. Figure 11-11 shows a pH meter with a mirrored scale.

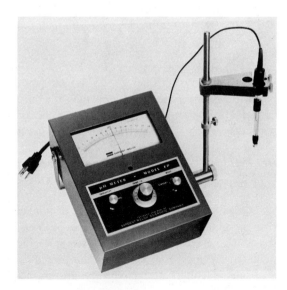

Figure 11-11. *pH meter using a mirrored scale to minimize parallax error. (Courtesy of Sargent-Welch Scientific Company)*

The other device for minimizing parallax is the *flattened pointer*. If we view a strip of paper edge-on, it appears very narrow. If we rotate it slightly one way or the other, it quickly appears to grow thicker. Using an instrument with a flattened pointer requires moving the head from side to side until a direction is found where the pointer appears its narrowest. When this happens, the flattened pointer is being viewed edge-on and the scale is being read from a perpendicular direction.

Although we have been talking primarily about the electrical meter movement, a pointer can be moved on a scale in other ways. There are, for instance, *pneumatic* indicators, in which air pressure transmitted through a piece of tubing actuates the movement. There are also *hydraulic* indicators, which use oil or water pressure to move the pointer. And, of course, there are strictly mechanical devices like clocks, planimeters, and bimetallic thermometers. What has been said about lag time, scale linearity, and parallax error applies to all these indicators.

Dead Zone

We should think about one more thing when using indicators with moving parts, including those with micrometer and vernier scales. Any mechanical device is made of parts that are subject to manufacturing tolerances. Furthermore, all moving parts are subject to friction. These two factors combine to produce a certain amount of "lost motion" when the instrument changes from one indication to another. You may notice this same effect with a car's steering mechanism. Even a new car has a certain amount of lost motion ("play") in the steering, but as a car gets older and the parts begin to wear, this lost motion increases. If you begin a turn in an old car, you may have to spin the steering wheel a noticeable amount before the front wheels begin to respond. So it is with mechanical indicators—the measured variable may have to change by a noticeable amount before the indicator begins to respond.

The lost motion in an indicator contributes to a *dead zone* in the instrument's response.

DEFINITION | An instrument's <u>dead zone</u> is the maximum variation in the measured variable that will produce no change in indication at all. The dead zone's size is determined by the amount of lost motion and friction in the indicator.

Indicators are usually manufactured to tolerances that keep the dead zone to less than 0.2 percent of the full-scale indication. However, as the mechanical parts wear and lost motion increases, the dead zone also increases. A large dead zone means a worn indicator.

Obviously, the dead zone contributes to systematic measurement error. Less obvious is the fact that this error can be positive or negative, depending on

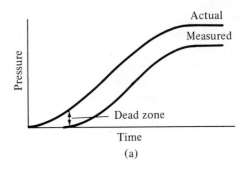

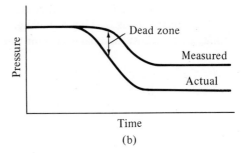

Figure 11-12. *Dead zone in a pressure indicator. The measured value is low when the pressure is rising (a), and high when the pressure is falling (b). If the pressure variation is completely within the dead zone (c), the indicator will not respond at all.*

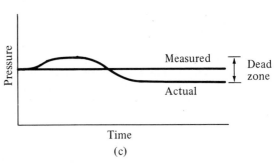

whether the measured variable is increasing or decreasing. The situation is shown in figure 11-12. The dead zone in this figure has been exaggerated to make it show clearly (this would be a *very* worn-out meter). In order not to confuse dead zone with lag time, the graphs in figure 11-12 assume essentially zero lag time. In part (a), we see what happens as the pressure (the measured variable) increases. The indicator does nothing at all until the pressure has changed a certain amount; it is this change that is called the dead zone. This initial change in pressure goes into taking up the lost motion in the indicator. As a result, the final reading is lower than the actual pressure by an amount approximately equal to the dead zone. In part (b), we see what happens when the pressure decreases, assuming that somehow the indicator happened to be reading correctly at the beginning. Again, the initial change in pressure takes up lost motion in the mechanism, so that the final reading is too high by an amount equal to the dead zone. In figure 11-12c, we see that the pressure may change up or down within the dead zone without having the change appear on the indicator at all.

Fortunately, there is a simple way to determine the dead zone on most indicators. Suppose that we suspect a voltmeter of being badly worn. We zero the meter, then measure the voltage of, say, a 6-V battery. Call this measurement V_{up}, since the pointer is moving up. Now measure the same voltage with the pointer moving down. This can be done by quickly switching the meter range so the pointer initially rises past the 6-V position, or by connecting the leads to a larger battery then quickly switching them to the first battery before the pointer drops very far. Call this measurement V_{down}. Then the dead zone is

$$\text{dead zone} = V_{down} - V_{up} \qquad (11\text{-}2)$$

How does dead zone differ from lag time? Lag time is an index of how long we have to wait for the pointer to reach a stable (and, it is hoped, accurate) reading. But no amount of waiting will encourage an indicator to take up its dead zone. Some high-quality instruments do have fairly large lag times, but no quality instrument has a large dead zone unless it is worn out.

Digital Indicators

Let's move on to some other types of indicators. One type of indicator that has become very popular in recent years is the *digital readout*. The idea, however, is old: to have a device that displays the actual numbers rather than requiring user interpolations from a scale. The early devices were purely mechanical: odometers, elapsed time indicators, and counters have long used little, numbered wheels that are rotated into position behind a window. Then came hot-filament tubes that light up a number by the flow of an electric current through a fine wire. These tubes tend to be fragile, but they do give a very large, bright display. Today we are all familiar with the LED (light-emitting diode) and liquid crystal displays that are used in hand-held calculators. These same digital readout devices appear in a broad range of instruments, some examples of which are shown in figure 11-13.

Cathode Ray Tube Indicators

Very often, the end product of a set of measurements is to be a graph. Yet if the measured variables change very rapidly, it may be impossible to make the required series of readings from a meter-type or digital indicator. In such cases the cathode ray tube (CRT) is a particularly useful indicator. This device is basically a variation of the television picture tube; it uses an electron beam to

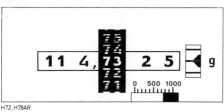

H72, H78AR

M11 × 72 × 30 mm   114,7325g

(a) An analytical balance with a mechanical display. Courtesy of Mettler Instrument Corporation.

(b) A frequency meter using hot-filament tubes. Courtesy of Heath Company.

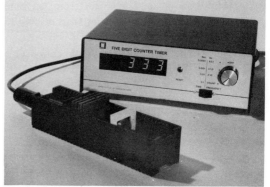

(c) A reflection densitometer using an LED display. Courtesy of Heidelberg Eastern, Inc.

(d) A counter/timer using a liquid crystal display. Courtesy of Daedalon Corporation.

Figure 11-13. *Digital displays are used in a wide variety of instruments.*

"write" on a phosphor-coated screen. The phosphor stays lit for a short time after the beam has passed—a few tenths of a second to a few seconds depending on the particular tube. The PPI radar indicator discussed in chapter 9 is a cathode ray tube designed to draw a polar graph. More commonly, the graph is Cartesian and the CRT indicator is part of an instrument known as an *oscilloscope*. Oscilloscopes come in many sizes and shapes; some are shown in figure 11-14. An oscilloscope can draw a graph of virtually any fast-changing variable versus time. Some can also be used to display several graphs at once, or to graph one variable versus another (cylinder pressure versus piston position in an engine, for instance). One limitation is that CRT graphs, or "traces," cannot be read very accurately because the CRT face itself is not very large. Another limitation is that the variables must change fairly rapidly because the persistence of the phosphor is never very long.

Strip Chart Recorders

Variables that change slowly are best recorded on a *strip chart recorder*. This device is basically a voltmeter, but again it can be used to indicate almost any physical variable through the use of an appropriate separate sensor. Chart recorders use a mechanically or pneumatically operated pen to draw a graph on a roll of graph paper. Generally, the rate at which the paper unrolls can be varied over a wide range, giving considerable control over the graph's time scale. The resulting graph is very accurate and has the obvious advantage of being a permanent record. Some strip chart recorders simultaneously graph several variables versus time. An example of this device is shown in figure 11-15.

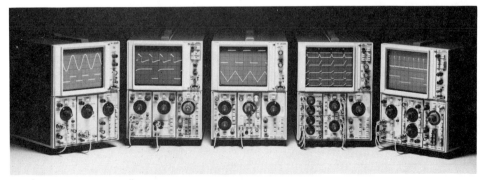

Figure 11-14. *Oscilloscopes are useful for graphing variables that change rapidly in time. (Courtesy of Tektronix, Inc.)*

Figure 11-15. *A strip chart recorder provides a permanent graph of a measured variable versus time. (Courtesy of Brookfield Engineering Laboratories, Inc.)*

Other Pen-type Recorders

Similar to a strip chart recorder is the X-Y recorder (figure 11-16). The difference is that an X-Y recorder graphs one variable versus another rather than a variable versus time. Again, any variable that can be converted to a voltage or pneumatic pressure can be plotted on an X-Y recorder.

Some recorders use slightly different arrangements for specific applications. We already saw a case of a polar coordinate recording unit on a psychrometer in chapter 9. Another familiar device is the drum-type recorder commonly used with barometers. Here the drum rotates once in a seven-day period, giving a week's record of temperature variations.

Null Measurement

So far, we have been discussing indicators that give a direct reading of the size of the measured variable. One important application of indicators, however, is

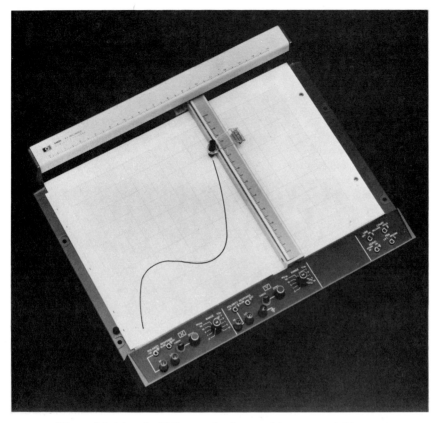

Figure 11-16. *An X-Y recorder for graphing one variable versus another. (Courtesy of Hewlett-Packard)*

in *null measurements*. In these cases the indicator's function is to give a *zero* reading when the measured quantity is balanced against a standard.

DEFINITION | A <u>null measurement</u> is made to detect the difference between the measured variable and a standard that it is being compared with.

The simplest example of a null measurement is the laboratory balance for determining mass. An unknown mass is placed on one side, while standard masses are added to the other side until the system balances mechanically. Sometimes the standard masses are built into the instrument, so they need only be slid back and forth rather than being lifted on and off. In any case, the idea is to find a *null point,* where the unknown is precisely balanced against the standards. This null point is indicated by a simple pointer on a scale. The divisions on this scale need not be calibrated in any special way, since only the zero point is important.

Figure 11-17. *An impedance bridge for measuring inductance, capacitance, resistance, and some related electrical quantities. The instrument functions on the principle of null measurement. (Courtesy of Hewlett-Packard)*

Although balances may look quite different, they all operate on this same principle.

A null measurement is the most accurate type of measurement. First, the standards are used directly in the measurement, rather than being used to calibrate an instrument that in turn is used to make the measurement. Second, the *difference* between two quantities usually can be indicated to a greater precision than the quantities themselves. We encountered this idea in chapter 4 when we talked about thermometers. A thermometer may have an accuracy of $\pm 1.0°C$, yet it may register temperature *differences* as small as $0.2\ C°$. With the balance, a small mass difference will swing the system one way or the other. Once a balance has been achieved, we can determine our unknown mass by totalling the standard masses it took to balance it.

Precise measurements of electrical capacitance, inductance, and resistance are often made with an impedance bridge. This device, shown in figure 11-17, is an electrical balance. The standards are internal to the instrument. The null indication is displayed by a conventional moving-coil meter movement. In a similar way, very accurate measurements of electric current can be made with a current balance, and voltage can be measured with a voltage balance known as a *potentiometer*.

In this chapter, we have seen that measurements may be read from a variety of indicating devices. Keep in mind, however, that regardless of the type of indicator and the variable being measured, the methods we developed earlier for the analysis of uncertainty and the description of the results still apply.

Summary

Every measuring instrument consists of a *sensor* and an *indicator*. Since a given sensor may often be connected to different indicators, or the same indicator to different sensors, it is convenient to talk about sensors and indicators separately.

The common types of indicators are the micrometer, the vernier, the dial (with its many variations), the mechanical and electronic digital indicators, the CRT, and the chart recorders. Each of these offers its special advantages and limitations; all convert information from a sensor to numerical values that the user can read. You should understand how the measurement result may be affected by *lag time, dead zone, parallax error,* and *scale nonlinearity* in the indicator.

When an indicator is used to show that the difference between two quantities is zero, it is called a *null indicator*. Such null measurements are more accurate than other types of measurement, since they involve direct comparisons with standards.

REVIEW QUESTIONS

1. What is a sensor?
2. What is an indicator, or display, unit?
3. Why is it convenient to describe instruments in terms of sensors and indicators?
4. What is the difference between a micrometer and a micrometer caliper?
5. What is a vernier?
6. Describe the operation of the d'Arsonval meter movement.
7. What is damping? Why is it necessary in moving-coil indicators?
8. What is lag time?
9. What is risetime?
10. What is the time constant of an instrument?
11. What is the approximate relationship between lag time and time constant?
12. Name some instruments that use the d'Arsonval meter movement as the indicator.
13. What is the difference between a linear and a nonlinear meter scale?
14. Name some instruments that typically have nonlinear scales.
15. What is parallax error?

16. Describe two techniques used to minimize parallax error.

17. Not all dial-type indicators are actuated by an electric current. What are some of the other principles that are used?

18. What is meant by "lost motion"?

19. What is a dead zone? How large is the dead zone in a high-quality dial-type indicator?

20. Describe the procedure for finding an indicator's dead zone.

21. Describe some digital indicators and their general principles of operation.

22. What is a CRT?

23. What is an oscilloscope? What kinds of variables can it display?

24. What is a strip chart recorder? What advantage does it offer over a dial-type indicator?

25. How does an X-Y recorder differ from a strip chart recorder?

26. Where might drum-type and polar recorders be used?

27. What is a null measurement? What is the advantage of such a measurement?

28. Name some instruments that use null measurement.

CHAPTER 12

Sensors

Separation of Sensors and Indicators

We have seen that most measuring instruments have a sensor and an indicator. While for very simple instruments the sensor and indicator are often combined in one unit, more commonly they are quite separate. Figure 12-1 shows a metallurgist measuring the temperature of a vat of molten iron. He does this by inserting a sensor (in this case, a thermocouple junction) into the liquid metal. The indicator, seen in the background, is a recording potentiometer. It must be far enough away so that the tremendous radiant heat does not damage its electronic components. A chart recorder is needed here because the thermocouple sensor is quickly destroyed by the molten iron, and the reading might be missed if it were not recorded automatically.

Sensor Design Requirements

A sensor, of course, is the part of an instrument that responds to changes in the variable being measured. At the same time, it must not respond to changes in other variables. If, for instance, a sensor is to detect temperature changes, it should not be sensitive to pressure, vibrations, stray electric and magnetic fields, air currents, or anything else. It should respond to temperature, and only temperature.

Furthermore, a given sensor will be useful for only a certain range of the measured variable. For instance, mercury freezes at -39°C, and its vapor pressure begins to get very high at temperatures around 350°C. This places a limit on the useful range of a mercury-filled thermometer. For lower and higher tempera-

Figure 12-1. *A metallurgist measures the temperature of a vat of molten iron using a thermocouple sensor. The indicator is a recording potentiometer, which gives a permanent record. The thermocouple itself is consumed during the measurement. (Courtesy of General Motors Corporation)*

tures, other types of sensors must be used. And even within this range, other sensors may offer advantages like ruggedness, sensitivity, and the ability for remote indication.

There are perhaps a hundred physical variables we could conceivably want to measure. Each of these variables requires different sensors for different ranges and different kinds of operating conditions. As a result, thousands of different sensors are manufactured, and it is impossible to discuss even a fraction of them here. What we will do, then, is to mention a few types of sensors for commonly measured variables, and to point out some of the principles of their use.

Sensor Capacitance

One characteristic property of sensors is their *capacitance*. We often think of capacitance as being strictly an electrical quantity, but in fact it can also be thermal, hydraulic, or mechanical.

> **DEFINITION** | *A sensor's* capacitance *is the ratio of the quantity of energy or material stored by the sensor to a unit change in the variable being sensed. In symbols,*

$$C = \frac{Q}{\Delta x} \qquad (12\text{-}1)$$

where C is the capacitance, Δx is the change in the measured variable, and Q is the quantity of energy or material the sensor must store to register this change.

Let's look at an example of how we might use this definition.

Calculating Sensor Capacitance

▶ **Example 12-1: The liquid-level gauge**
A simple liquid-level gauge is shown in figure 12-2. In the glass tube, which is the sensor, the liquid seeks the same level it has in the closed tank. The indicator is a ruler that allows the liquid level to be read. Let's suppose that the tank is 30.0 cm in diameter, while the glass tube has an inside diameter of 2.00 cm. To sense the liquid level, the gauge draws off some of the liquid in the tank. If the level in the tank (the variable being measured) rises 1 cm, the tube must store an additional volume of liquid

$$Q = \pi r^2 h \quad \text{(formula for the volume of cylinder)}$$
$$= \pi (1.00 \text{ cm})^2 (1 \text{ cm})$$
$$Q = 3.14 \text{ cm}^3$$

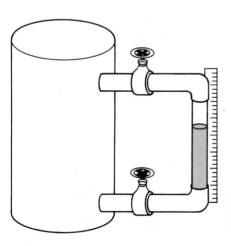

Figure 12-2. *A liquid-level gauge on a storage tank.*

The capacitance is therefore

$$C = \frac{Q}{\Delta x} \qquad (12\text{-}1)$$

$$= \frac{3.14 \text{ cm}^3}{1 \text{ cm}}$$

$$C = 3.14 \text{ cm}^3 \text{ per centimetre}$$

In other words, if we pour more liquid into the tank, 3.14 cm³ goes into the gauge for each 1-cm increase in the level. Notice that the capacitance of the gauge here has nothing to do with the size of the tank it is connected to. ◄

The Uncertainty Principle

We can appreciate the importance of capacitance if we think about this: The tank in figure 12-2 has two stopcocks to isolate the level gauge. Suppose we begin with an empty tank, close the stopcocks, then pour some liquid into the tank. If we now want to know what the level is, we open the stopcocks and allow the gauge to fill. Do we get a true reading of what the level was before the stopcocks were opened? No. The level in the tank has to drop slightly to fill the gauge, and it is this *final level* that is indicated. In other words, the gauge has affected the variable that it is supposed to measure.

Now if the gauge's capacitance is large but the tank is small, the gauge will have a very large effect on the liquid level in the tank. If, on the other hand, the gauge's capacitance is small compared with the size of the tank, then this effect will be small. It is a basic principle of measurement that sensors tend to affect the quantity they are sensing. This principle is called the *uncertainty principle*.

DEFINITION | *The uncertainty principle states that introducing a sensor into a system will cause a change in the variable that the sensor responds to. The size of this effect is determined by the sensor's capacitance.*

Let's consider the micrometer caliper. This is a very precise instrument, and it may seem that it has no capacitance. Yet the caliper senses the size of the test sample by applying a small stress through the spindle and anvil—stress that is regulated by the torque on the ratchet. The ratchet slips when the sample is compressed to the point where its elastic counterstress balances the applied stress. This is the "Catch-22" of measurement: the caliper *cannot* sense the

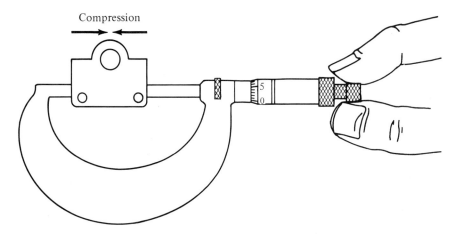

Figure 12-3. *Capacitance in a micrometer caliper. To sense the size of the sample, the caliper spindle must compress the sample a very small amount. This places a limit on the accuracy of the instrument.*

sample's size unless there is this contact pressure, yet the pressure slightly alters the size of the sample. The effect is shown in figure 12-3. Thus the micrometer caliper does have some capacitance, and because of this capacitance, even vernier micrometers cannot measure more accurately than ±0.005 mm.

Theoretically, all sensors have capacitance. In some cases, however, this capacitance is so small that its effect is overshadowed by errors of indication. Many optical sensors fall into this category. In such cases we may act as though the sensor has no capacitance at all. The *infrared pyrometer,* for instance, is used to measure temperature. The instrument intercepts thermal radiation from a hot sample, and the characteristics of this radiation form the basis of the temperature indication. The sensor does not actually touch the sample. The only interaction is that the sensor (and instrument) reflect some of the radiation back toward the hot body, and therefore affect its temperature in a very small way. Since the indicator has a typical accuracy of ±6 C° or less, the small sensor capacitance in this case is of no practical consequence.

In general, however, we have to expect that thermometers cause changes in temperature, that pressure gauges affect pressure, that voltmeters alter voltage, that ammeters restrict current, that flowmeters affect fluid flow, that hardness testers change a metal's hardness, and so on. Although we cannot say that a clock alters the time it measures, a mechanical clock escapement does affect the rate of oscillation of the clock's pendulum or balance wheel. Similarly, a crystal oscillator has a slightly different frequency when it oscillates alone or when it is used to regulate an electronic clock.

Capacitance and Lag Time

A sensor's capacitance is related to another quantity discussed earlier: the instrument's *lag time,* or *time constant.* In the last chapter, we attributed lag time to the need for damping in the indicator. Actually, this was only half the story. Lag time is also affected by the sensor's capacitance. Sensors with large capacitance tend to introduce large lag times in the indication.

We can see why this is so if we examine figure 12-4, which is a diagram of a resistance thermometer. The sensor is an electrical resistor whose resistance is affected by temperature. When the sensor is placed into the liquid, two things happen. First, some of the heat from the liquid goes into heating the sensor. This causes a slight drop in the liquid's temperature, just as placing a cold spoon in a hot cup of coffee cools the coffee slightly. Second, the sensor does not come to the temperature of the sample *immediately,* but both the sensor and the sample move toward a common equilibrium temperature. This is because it takes time for heat to flow from the sample into the sensor. As the sensor's temperature rises, the indicator continuously displays a rising temperature. To complicate matters further, this indication lags the actual sensor temperature because of the indicator's own time constant.

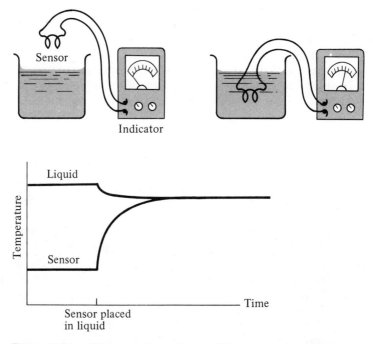

Figure 12-4. *Effects of thermal capacitance in a resistance thermometer. The sensor's capacitance causes a change in the temperature being measured, as well as a time lag in the indication.*

We can estimate the drop in the sample's temperature mathematically if we know the sensor's thermal capacitance and if we are clever enough. Such calculations are too specialized to deal with here; the point is that they *can* be done. If there is any choice in the matter, we are obviously better off with a sensor that has a small rather than a large capacitance. Low-capacitance sensors offer two advantages: they have little effect on the variable being measured, and they make little contribution to the indicator's lag time.

Let's look at just a few of the sensors we can use for measuring some different variables.

Measuring Length and Distance

The most frequent length and distance measurements are made with very common instruments: rules, tape measures, micrometer and vernier calipers, and so on. We also discussed radar distance-measuring equipment in chapter 9. To this list, we may add a few other interesting instruments, all of which have very low capacitance.

The *odometer* is the familiar mileage indicator on the dashboard of a car. It is basically a mechanical counter that keeps track of the revolutions of one of the car's wheels and converts this measurement to distance. Because a tire's circumference expands at high speeds and decreases as the tire wears, this device has an accuracy of only around ±5%. A similar device is the *revolution counter* used for estimating distances on a map. Because roads are seldom straight, it is impossible to use a conventional rule for this type of measurement. The revolution counter allows the exact path of the road to be traced, and the indicator usually gives the map distance in inches or centimetres. Referring this measurement to the map scale allows the distance in miles or kilometres to be estimated. With care, an accuracy of ±3% or better can be achieved.

Ultrasonic depth meters use the reflection of sound from an underwater object to determine the object's distance from a sensor. The sensor, called a *transducer,* introduces a high-frequency (ultrasonic) sound wave into the water and receives any reflected signal. The time difference between the outgoing and incoming signals is then converted to a distance indication. Figure 12-5 shows the special recording indicator used with one such device; the chart shows the profile of the bottom of the body of water. A device called an *ultrasonic thickness gauge* uses the same principle to measure the wall thickness of a piece of pipe without cutting the pipe open. The sensor is still a transducer, and the indicator is often an oscilloscope.

The *cathetometer* is a precision instrument for the remote measurement of vertical distances. The object is sighted through a carefully leveled telescope that has a fine cross hair in the field of view. The telescope can be cranked up and down, and the distance the cross hair moves is read on a vernier scale. The instrument has a precision of 0.05 mm and a comparable accuracy. The *traveling*

Figure 12-5. *An ultrasonic depth gauge. (Courtesy of Heath Company)*

microscope uses the same principle in measuring very small objects. A cross hair in the microscope eyepiece is lined up with one end of the object to be measured, then the microscope is moved to where the cross hair lines up with the other end of the object. A vernier or micrometer scale indicates the distance traveled. This technique offers the advantage of avoiding physical contact with the sample. In both the cathetometer and the traveling microscope, the sensor is the cross hair and the optical system, used in conjunction with the human eye.

A *laser* is a light source with a single, well-defined wavelength and a very narrow angle of divergence. A beam of laser light spreads out so little that it has been possible to aim one at the moon and to intercept the reflection. From this, the distance to the moon has been measured to an accuracy of a fraction of 1 m. Very small distances, on the order of 10^{-8} m, have been measured with a laser in conjunction with an *interferometer*. The laser also has been used to measure the "creep" along geological fault lines by aiming the beam across the fault, reflecting it back with a mirror, and monitoring the movement of the reflected beam from week to week.

Strain is the deformation experienced by a sample subjected to mechanical stress. In most technical applications, strain is a very small yet very important quantity, generally measured by *strain gauge* sensors whose electrical properties change when they are stretched or compressed. Figure 12-6 shows strain measurements being made on a scale model of a car body.

Measuring Speed and Frequency

The most accurate speed measurements are made by measuring distance and time separately, then performing a calculation such as the one in example 7-6. More direct methods sacrifice accuracy for the sake of convenience.

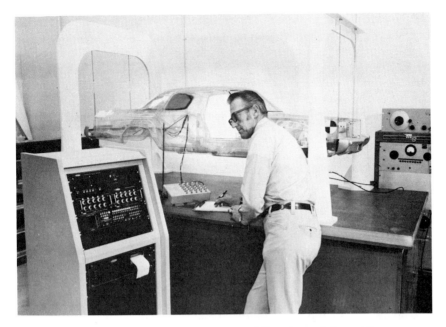

Figure 12-6. *Taking strain data in a 3/8 scale-model car. (Courtesy of Chrysler Corporation)*

The automobile *speedometer* is driven by a flexible cable connected to the vehicle's front wheel or transmission. This cable rotates a permanent magnet within a "speed cup" mounted on a spindle. The speed cup is held in position by a hairspring. The rotating magnetic field transmits a torque to the speed cup, which twists to the point where this torque is counterbalanced by the hairspring's torque. The device operates much like the d'Arsonval meter, except that the magnet is constantly rotating and the pointer is connected to the separate speed cup. The speedometer has a moderate capacitance, due to the rotating magnet's inertia, and therefore it does not lend itself to use on small moving objects.

The *stroboscope* measures rate of rotation, commonly expressed in revolutions per minute. An electronic circuit fires a xenon flash tube at an adjustable rate. When the flash rate is synchronized with the rate of rotation of a fan or motor, the rotating part appears to be standing still. One difficulty is that the sample will also appear to stand still if the flash rate is half, one-third, or some other integer fraction of the actual rotation rate. For this reason, the user must have some skill to use the instrument properly.

The *tachometer* also measures frequency of rotation. The most common version uses a momentary-contact switch that briefly closes once during each revolution of the rotating part. This action allows a current to charge a capacitor, which is subsequently discharged through a d'Arsonval movement or other suitable indicator. The meter indicates the average current, which depends on the frequency of the interruption. Another version uses a small direct-current electric generator as the sensor. The generator is driven by the rotating part, and it produces a voltage that increases with the frequency of rotation.

The *pneumatic speed transmitter* is basically a precision valve that is driven by the rotating part. A constant-pressure air supply is connected to the valve's input, and the output pressure varies with the frequency of rotation. This output air pressure may be transmitted up to several hundred feet through a piece of pneumatic tubing. The measurement is displayed on a pneumatic indicator that responds to changes in air pressure in the connecting line.

Measuring Pressure

Since pressure is such an important variable in industrial processes, hundreds of different sensors have been designed for its measurement. Some of these sensors respond to the *difference* between the measured pressure and the prevailing atmospheric pressure. This difference is called the *gauge pressure*. A gauge pressure sensor registers zero pressure when it is open to the atmosphere. Other sensors respond to *absolute pressure*. These sensors register the actual value of the atmospheric pressure when they are open to the atmosphere, and register zero only when exposed to a very high vacuum. The relationship between these pressures is

absolute pressure = gauge pressure + atmospheric pressure

The "standard" value of atmospheric pressure is $1.013\ 25 \times 10^5$ Pa, or $14.695\ 95$ lb/in^2. However, the actual value varies considerably depending on location and weather conditions. Consequently, it is not so simple to convert an absolute pressure to a gauge pressure unless a separate measurement of the atmospheric pressure is available. It is much better to select an appropriate sensor carefully to begin with. Figure 12-7 shows the maximum useful range for some selected types of pressure sensors.

An absolute pressure gauge whose useful range is mainly below atmospheric pressure is called a *vacuum gauge*. *Thermal vacuum gauges* operate on the principle that the rate of heat loss from a hot wire depends on the air pressure. *Ionization vacuum gauges* are electron tubes that sense the rate of production of gas ions by a small electron current. They are useful only for high vacuums because the filament will burn out quickly if the pressure is too great.

Mercury barometers and *manometers* sense pressure by the height of a column of mercury that is supported by pressure to be measured. *Spiral, helix,* and *Bourdon* tube sensors consist of a closed section of coiled or partly coiled tubing that tends to uncoil when the inside pressure increases. These sensors may be linked mechanically to an indicator, or they may be used to vary the resistance or inductance of a coil to give an electrical output. The *diaphragm capsule* is a series of circular, hollow, metal diaphragms, with one end closed and the other end open to the pressure to be measured. An increase in pressure

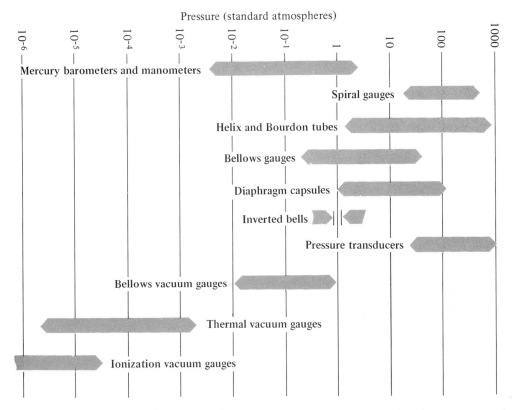

Figure 12-7. *Useful range of various pressure sensors.*

causes a distortion of the whole assembly and a slight movement at the closed end. Again, this movement may be linked mechanically to a pointer, or used to give an electrical output. *Bellows gauges* operate similarly, except that the bellows are more flexible and require a separate external spring to expand against. *Pressure transducers* convert high pressures directly to electrical voltage. *Inverted bell sensors* are very sensitive devices for detecting small pressure changes near atmospheric pressure. The inverted bell floats in a bath of oil, and the flotation level varies with the pressure. Several of these pressure sensors are shown in figure 12-8.

Measuring Temperature

Of all the physical variables, temperature is the most difficult to measure accurately and precisely. At temperatures near the triple point of water (0.01°C), the limit of sensitivity of the best commercial instruments is around 0.001 C°. At other temperatures, uncertainties are quite a bit higher. Table 12-1 lists the

(a) Bellows type. Courtesy of Fischer & Porter Company.

(b) Bourdon tube. Courtesy of Honeywell Inc.

(c) Diaphragm capsule. Courtesy of Honeywell Inc.

(d) Pressure transducer. Courtesy of Honeywell Inc.

Figure 12-8. *Some examples of pressure sensors.*

limits of accuracy, both for the best available commercial instruments and for a standards laboratory using extensive calibrations and cross-calibrations. Many common instruments will not come close to these figures—a mercury-filled thermometer, for instance, has a limit of around ±0.1 C°, and the bimetallic indoor-outdoor thermometers may be good to only ±1.0 C°. Figure 12-9 shows the useful range for some different temperature sensors.

The *gas thermometer* is a very accurate sensor for low temperatures, but it has a high capacitance and is fragile. It functions on the principle that the pressure of a confined gas—usually helium, because of its low boiling point—depends on the gas's temperature. The pressure is usually read from a manometer. The *ultrasonic thermometer* measures the velocity at which high-frequency sound travels through a substance. This velocity gives a measure of the temperature. The *spectrometer* is used as a temperature sensor in astronomical work, where the properties of a star may be of interest. It also finds limited use as a thermometer in certain specialized research areas. Basically, however, it remains

Table 12-1. *Limits of Precision in Temperature Measurement*

	Uncertainty (C°)	
Temperature	Practical Limit	Limit of High-quality Commercial Instruments
15K	±0.05	>±0.05
60K	0.02	>±0.02
90K	0.005	0.01
0.01°C	0.000 3	0.001
200°C	0.002	0.004
400°C	0.002	0.004
630°C	0.3	3
1 100°C	0.3	5
1 800°C	3	10
2 800°C	7	25
3 500°C	13	50

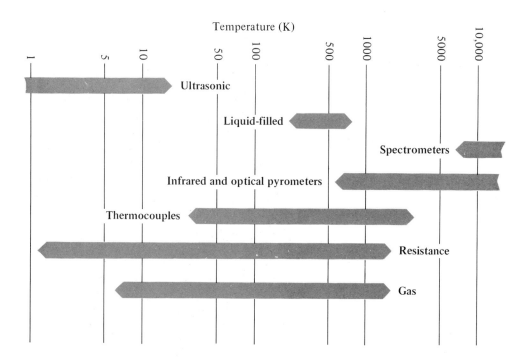

Figure 12-9. *Useful ranges of various temperature sensors.*

Sensors / 221

a device for measuring wavelengths of light. The other temperature sensors listed in figure 12-9 have already been discussed elsewhere.

Chemical Sensors

Analyzing the chemical composition of a sample can be a time-consuming laboratory procedure if the analysis must begin from scratch. Often, however, certain possibilities can be excluded at the outset. A metallurgical sample, for instance, is not likely to contain organic compounds, nor is a fuel gas likely to contain metal ions. Instrumentation is available for quick detection and measurement of certain elements and simple compounds, but it is necessary to know what you are looking for.

A basic version of the *spectrometer,* already mentioned briefly, is shown in figure 12-10. The material to be analyzed is burned and ionized in a carbon arc. The resulting light is allowed to enter the collimating tube, where the light is confined by an adjustable slit to a narrow, well-defined beam. This beam is directed through a glass prism, or, more commonly, a *diffraction grating.* This grating is a piece of glass ruled with a large number of closely spaced parallel

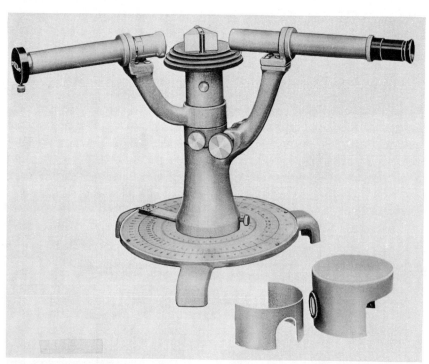

Figure 12-10. *An optical spectrometer. (Courtesy of Sargent-Welch Scientific Company)*

scratches or "lines"—as many as 6 000 per centimetre. The prism or grating separates the light into its separate wavelength components, sending each component in a different direction. These different components appear as different colors, and since each is an image of the collimator slit, they are also called "lines." (Photographs of two line spectra are shown in figure 8-8.) An angular scale can be read to get the direction of each line, and a mathematical formula is used to calculate each wavelength. For analytical purposes, these wavelengths need not be measured very accurately. Each chemical element has its own characteristic line spectrum, which has been well documented in various handbooks. The existence of a certain sequence of wavelengths therefore indicates the presence of a particular element. The procedure may be refined so that the brightness, or relative intensity, of the lines gives a measure of the relative abundance of each of several chemical elements in the sample. An instrument that does this is called a *spectrophotometer.*

Many organic chemical compounds selectively absorb specific wavelengths or bands of wavelengths of light. These compounds may be detected by an *absorption spectrophotometer.* The sample may be prepared as a solution through which light of varying wavelength is projected. The wavelengths used may extend into the ultraviolet and infrared regions. A *photoelectric sensor* gives a measure of the percent of the original light at a particular wavelength that is transmitted by the sample. A graph of this percent of transmission versus wavelength is the sample's *absorption spectrum.* Comparing this spectrum with standard absorption spectra leads to an identification of the material. It is also possible to determine the relative proportions of two or three different chemicals through a mathematical analysis of the composite absorption spectrum.

Exhaust gases may be analyzed for the presence of carbon monoxide, nitrous oxides, sulfur oxides, and unburned hydrocarbons. This is often done by a procedure known as *gas chromatography,* where the various components are separated and identified by their physical properties. The technique involves a considerable laboratory setup. A more direct procedure is the use of *electrochemical sensors,* which are usually sensitive to a particular gas. Such a sensor is shown in figure 12-11, where it is being installed directly in the exhaust system of an experimental engine. The output from this particular sensor will be used to control the engine's fuel-air mixture.

Radiation Sensors

Ionizing radiation may consist of alpha particles, beta particles, gamma rays, x rays, or a combination of these forms. Alpha particles are heavier and more highly charged than the others; as a result, they are not very penetrating and not very useful (although swallowing an alpha-radioactive substance would be lethal). Radiation sensors exploit ionizing radiation's ability to produce electrically charged atoms (ions) and to induce chemical reactions.

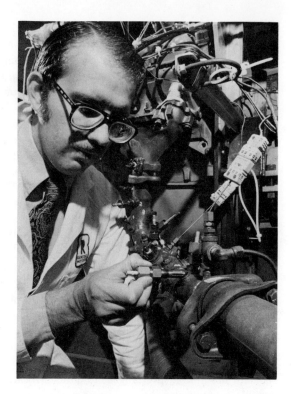

Figure 12-11. *An electrochemical sensor used to continuously monitor engine exhaust gases. (Courtesy of General Motors Corporation)*

The simplest sensor is a *dosimeter,* which consists of a piece of photographic emulsion sealed in a small light-tight envelope. When developed under controlled conditions, the negative's density is a measure of the accumulated radiation exposure, or "dose." Dosimeters are commonly worn by personnel who work in areas subject to radiation hazards.

Ionization chambers and *Geiger-Müller tubes* operate similarly. A glass tube contains a gas under a very low pressure and two separate electrodes with a voltage maintained across them. When an ionizing particle enters the gas, the resulting charged atoms move toward the negative electrode and the free electrons move toward the positive (high-voltage) electrode. A small current momentarily flows in the external circuit, and this pulse is counted by an appropriate indicator. A *proportional counter* is a specially designed ionization chamber that allows each beta or gamma ray to produce a voltage pulse proportional to its energy. *Semiconductor junction sensors* are the solid-state counterparts of these devices. The *scintillation counter* uses a phosphor to detect the radioactive particles. Each intercepted particle produces a flash of light, which is converted to an electrical signal by a *photomultiplier tube.*

Calorimeters, which are commonly used to measure quantities of heat, may also be used as radiation sensors. When radiation is absorbed, the energy is eventually downgraded to heat. A measurement of this heat therefore gives the energy in the absorbed radiation.

Cloud chambers work on the principle that an ionizing particle will leave a trace of its path in a container of air saturated with water vapor. These devices are useful for monitoring the behavior and interactions of small numbers of ionizing particles. A similar device is the *bubble chamber,* which uses a liquid saturated with a dissolved gas—similar to a glass of beer saturated with carbon dioxide. Each ionizing particle leaves a trace of bubbles in its path.

Radiation sensors find use in varied fields. Archaeological finds can be dated by carefully measuring their beta radioactivity. X rays can be used to measure the thickness of pipes or to detect flaws in welds and castings. The *beta gauge* shown in figure 12-12 is being used to continuously monitor the density of paper being produced at a paper mill. The paper may be moving at speeds as high as 1 000 m/min. Without this technique, samples of the paper must be cut and weighed individually. *Radioactive tracers* are radioactive chemicals that can

Figure 12-12. *Measuring the weight density of paper at a paper mill: (a) individual samples being weighed on a balance; (b) continuous measurement using a beta gauge. (Courtesy of Union Camp Corporation)*

Sensors / 225

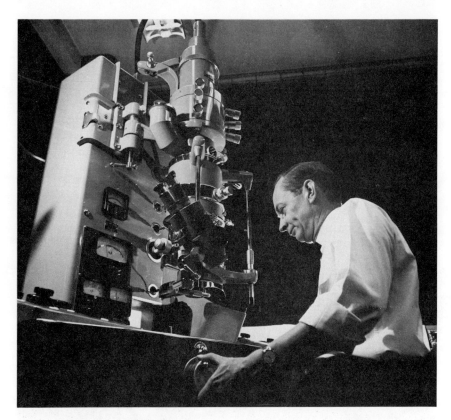

Figure 12-13. *An electron microscope. (Courtesy of Westinghouse Electric Corporation, Defense Group)*

be introduced into a system to monitor such diverse things as the function of the thyroid gland or the rate of wear on piston rings. Radioactive sources are also used in certain types of vacuum gauges, where the rate of ionization gives an indication of the air pressure.

Photographs as Sensors

Sensors that operate on the principle of photochemical sensitivity are a special, important category. There is little point in arguing whether a photograph is actually a sensor or an indicator, for in some cases it is both. If we use a hidden camera to photograph a bank hold-up, we have not only sensed (or proved) that the hold-up has taken place, but we have also indicated and recorded such information as the physical description of the culprits. In technical applications, we usually use photographs as an intermediate step in a measurement process.

For instance, we may photograph a trace on an oscilloscope screen, then later measure the characteristics of the trace from the photograph. Or we may photograph the spectrum of a distant star, then analyze the star chemically by measuring the densities of the spectral lines from the negative.

Photographic film is useful for sensing spatial intensity patterns in light, ionizing radiation, ultraviolet and infrared radiation, and electron beams. Figure 12-13 shows an *electron microscope,* a device used to magnify small objects by 200 000 diameters or more. (The conventional optical microscope has about a 2 000-diameter limit.) The sample is placed in a vacuum chamber where a beam of high-energy electrons is passed through it. Magnetic lenses focus and enlarge the resulting beam pattern, then a photographic plate both senses and records the magnified image. Besides its obvious applications in microbiology and surface chemistry, the procedure can yield important information about the basic structure of matter.

Figure 12-14 shows *electron diffraction* patterns obtained from a sample of a crystalline material. This photograph is not the image of the atoms themselves, but of patterns produced by the diffraction of electrons as they pass through the material's lattice structure. The spacing of the spots and rings can

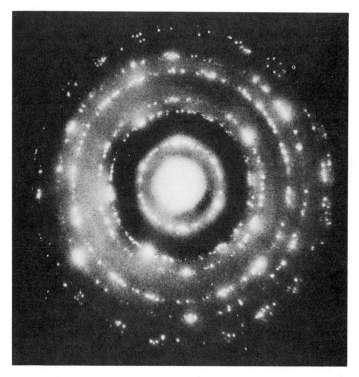

Figure 12-14. *An electron diffraction photograph of polycrystalline hexagonal P-graphite. (Courtesy of Sargent-Welch Scientific Company)*

be measured from the photograph with a vernier caliper or traveling microscope, and analyzed mathematically to determine the sample's actual structure.

Another interesting application of photography is NASA's LANDSAT system. LANDSAT is a satellite that circles the globe 14 times each day, continuously transmitting electronically coded information about the terrain it passes over. The receiver on the ground automatically converts this information to photographs such as the one shown in figure 12-15. Such photographs can be used for monitoring meteorological conditions, the growth or shrinkage of glaciers, forest fires, water impoundments, strip-mining, clear-cutting in forested regions, and agricultural productivity. The images are free of most of the dis-

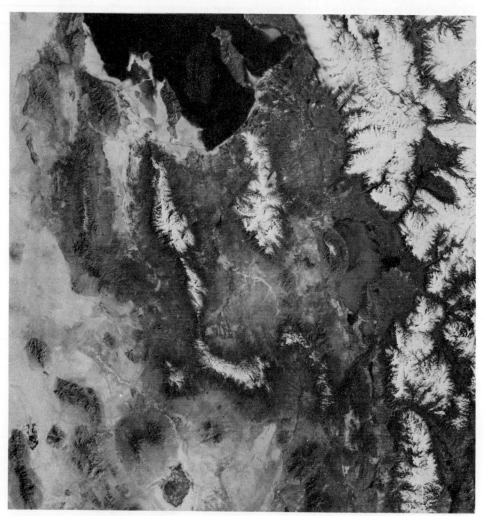

Figure 12-15. *LANDSAT photograph of Salt Lake City and surrounding area. (Courtesy of National Aeronautics and Space Administration)*

tortion found in lower-altitude aerial photographs, and they are much cheaper to make now that the satellite is already in orbit.

Summary

A *sensor* must respond to changes in the variable of interest, while remaining unaffected by changes in other variables. With a given sensor, this requirement will be met only for a certain range of the measured variable; therefore different sensors must be used for different ranges. Furthermore, some sensors offer special advantages like ruggedness, ability for remote indication, low cost, and so on. As a result, there are many, many different kinds of sensors. The few mentioned in this chapter should give some idea of the variety.

Although choosing the best sensor for the particular application is not always easy, the instrument manufacturers themselves tend to be very helpful if they are contacted. It is, after all, not to their advantage for their products to be misused and thus produce poor results. For this reason, most instrument manufacturers distribute much technical information and laboratory data that cannot be found in books or periodicals. Such information is kept on file at most laboratories (both college and industrial).

REVIEW QUESTIONS

1. Explain why many types of sensors are manufactured for measuring the same physical quantity.
2. What is a sensor's capacitance?
3. What is the practical consequence of a high sensor capacitance? a low sensor capacitance?
4. What is the uncertainty principle?
5. Give an example of a sensor whose capacitance is so low that its effect on the measurement may be neglected.
6. Give an example of a sensor whose capacitance is high enough to clearly affect the measurement.
7. How is sensor capacitance related to instrument lag time?
8. Describe six different techniques for length measurement.
9. What is strain? How might it be measured?
10. What is the most accurate way to measure speed?

11. What is a stroboscope?
12. What does a tachometer measure?
13. What is a pneumatic speed transmitter?
14. What is gauge pressure?
15. What is absolute pressure?
16. What is a vacuum gauge?
17. Why can an ionization-type vacuum gauge be used only for very high vacuums?
18. Describe the operation of five different types of pressure sensors.
19. Describe the operation of three different types of temperature sensors.
20. How may a spectrometer be used for chemical analysis?
21. What is a diffraction grating?
22. What is a spectrophotometer?
23. What is a particulate sampler used for?
24. Describe two instances where chemical sensors are used.
25. Describe the operation of five different types of radiation sensors.
26. Describe two applications of photographic sensors.
27. What is an electron microscope?
28. What is an electron diffraction pattern?

CHAPTER 13

Dimensional Analysis and the International System

This book began by pointing out that a measurement result consists of two things: a number and a unit. So far, most of our attention has been devoted to dealing with the numerical part. The unit, however, is equally important, and in this chapter we will look at it more closely.

We have seen that units are defined in terms of physical standards. Standards come in many sizes and shapes, and some are better than others. All standards, however, are related through measurements to the seven International System (SI) primary standards described in chapter 1. A slightly more technical description of these same standards is given in appendix table B-1.

Multiples of Base Units

One of these seven SI primary standards defines the unit of length known as the *metre*. (We use the French spelling here to avoid confusion with the *meter* that is a measuring instrument.) This is the only unit of length in the International System. Yet nearly 50 other units of length not part of the SI are still in use. At one time, each of these different length units had its own primary standard, some as unscientific as a number of barleycorns placed end-to-end. Since such standards tended to be very complicated, confusing, and often inconsistent, governments began to redefine some of these units in terms of other units. A foot, for instance, came to be defined as one-third of a yard, and an inch as one-

Table 13-1. *The Seven Fundamental Physical Quantities and Their SI Units*

Quantity	SI Unit	Symbol
length	metre	m
mass	kilogram	kg
time	second	s
electric current	ampere	A
temperature	kelvin	K
luminous intensity	candela	cd
amount of substance	mole	mol

twelfth of a foot. This meant that the standard for the yard could also serve as the standard for the foot and the inch. When the original metric system was developed in France in 1789, definitions like these were built right in. Centimetres, decimetres, dekametres, kilometres—all distance units were defined as a power-of-ten multiple of the base unit, the metre. In the metric system, then, only one length standard was needed from the very beginning.

Furthermore, a simple set of rules was developed for naming multiples of base units. These rules are summarized in appendix table B-4, which lists the names of prefixes that are used to designate various power-of-ten multiples of SI units. An ampere, for instance, is a rather large unit for describing many electric currents; a submultiple that is only one one-thousandth as large is the *milli*ampere (mA). An even smaller submultiple, one one-millionth of an ampere, is the *micro*ampere (μA). Very large currents may be expressed in kiloamperes (kA), just as large distances may be expressed in kilometres (km); the prefix *kilo* designates a multiple of 1 000, and so on. The *kilogram* is the only SI base unit that, for historical reasons, already has a prefix.

Today's International System has seven base units and seven primary standards because there are only seven different kinds of physical quantity we can measure. All other physical quantities are some combination of these basic seven. It is therefore essential that these seven be committed to memory. They are listed again in table 13-1.

Unit Conversions

Often a measurement must be converted from one unit to another. A measurement made in kilograms may be needed to be expressed in pound-mass (lbm), or a number of yards may have to be converted to an equivalent number of metres. Obviously, it is impossible to convert a number of feet (distance) to metres per second (speed), or kilograms (mass) to newtons (force). Conversions can be made only between units for the same physical quantity. The mathematical procedure for doing such conversions is based on treating units as algebraic quantities. For example, suppose that a distance x has been measured as

$$x = 17.341 \pm 0.003 \text{ in.}$$

The unit here is the *inch,* and it has been abbreviated *in.* in the standard way.* We may need to represent this result in units of centimetres (cm). To do so, we obviously need some information relating the size of these two units. It turns out that the relationship is

$$1 \text{ in.} = 2.54 \text{ cm}$$

This happens to be an exact relationship, established by definition. The algebraic consequence of this equation is that

$$\frac{1 \text{ in.}}{2.54 \text{ cm}} = 1$$

and that

$$\frac{2.54 \text{ cm}}{1 \text{ in.}} = 1$$

This is just the familiar fact that dividing a quantity by an equal quantity gives a quotient of 1, regardless of the way the division is performed.

We may now set up the conversion as follows:

$$x = (17.341 \text{ in.} \pm 0.003 \text{ in.}) \left(\frac{2.54 \text{ cm}}{1 \text{ in.}} \right)$$

All we are doing here is multiplying the original result by 1, which of course leaves it unchanged. However, we have written the converting fraction so that the units *in.* cancel algebraically and the result is expressed in *cm*. This gives

$$x = (17.341 \text{ in.}) \left(\frac{2.54 \text{ cm}}{1 \text{ in.}} \right) \pm (0.003 \text{ in.}) \left(\frac{2.54 \text{ cm}}{1 \text{ in.}} \right)$$

$$x = 44.046 \text{ cm} \pm 0.008 \text{ cm}$$

Originally, our distance was being compared to a length unit called the "inch"; now it has been compared to a unit called a "centimetre." Keep in mind that the physical distance itself is not changed by this mathematical procedure. The following example summarizes the calculation.

---

*The abbreviation for the inch is the only one that is followed by a period, to avoid confusion with the word *in.*

Conversion of Simple Units

▶ **Example 13-1: Gallons and liters**
The volume capacity of a certain tank is determined to be

$$V = 20\ 318 \pm 25 \text{ gallons (gal)}$$

This result is to be converted to liters. To do this, we need to know that

$$1 \text{ gal} = 3.785\ 412 \text{ ℓ}$$

(This is the USA liquid gallon, which is different from the USA dry gallon and the old Canadian and United Kingdom liquid gallons.) Once we have this relation, we can write

$$V = (20\ 318 \text{ gal} \pm 25 \text{ gal})\left(\frac{3.785\ 412 \text{ ℓ}}{1 \text{ gal}}\right)$$

$$= (20\ 318 \text{ gal})\left(\frac{3.785\ 412 \text{ ℓ}}{1 \text{ gal}}\right) \pm (25 \text{ gal})\left(\frac{3.785\ 412 \text{ ℓ}}{1 \text{ gal}}\right)$$

$$V = 76\ 912 \text{ ℓ} \pm 95 \text{ ℓ}$$

Although the conversion factor itself is accurate to seven significant digits, the result has been expressed to the same significance as the original number. ◀

About this time, someone always objects that writing the procedure algebraically is more trouble than it is worth. The following example should dispel such objections.

Conversion of Compound Units

▶ **Example 13-2: Miles per hour to metres per second**
A certain speed is measured in units of miles per hour, with the result

$$v = 6\bar{0}\ \frac{\text{mi}}{\text{h}} \pm 2\ \frac{\text{mi}}{\text{h}}$$

This is to be expressed in metres per second.

Since speed is distance divided by time, the unit of speed is written algebraically as a unit of distance divided by a unit of time. In this scheme of things, then, we write mi/h rather than mph. We will be converting to units of m/s. To do so, we need to call on three conversion relationships:

$$1 \text{ mi} = 1\ 609.344 \text{ m}$$
$$1 \text{ h} = 60 \text{ min}$$
$$1 \text{ min} = 60 \text{ s}$$

Using these relationships, we can write the entire conversion in one equation:

$$v = \left(60\,\frac{mi}{h} \pm 2\,\frac{mi}{h}\right)\left(\frac{1\,609.344\,m}{1\,mi}\right)\left(\frac{1\,hr}{60\,min}\right)\left(\frac{1\,min}{60\,s}\right)$$

Remember that there are two ways to write each converting fraction. We choose the one that allows the unwanted units to cancel algebraically. The unit *mi* in the original numerator must be canceled with a *mi* in a denominator, and the unit *h* in the original denominator must be canceled with an *h* in a numerator. When this happens, we know we have set up the conversion correctly. Then

$$v = \frac{(60)(1\,609.344)}{(60)(60)}\,\frac{m}{s} \pm \frac{(2)(1\,609.344)}{(60)(60)}\,\frac{m}{s}$$

$$v = 26.8\,\frac{m}{s} \pm 0.9\,\frac{m}{s}$$

Thus a vehicle that is cruising at $6\bar{0}$ mi/h is traveling 26.8 m each second. ◀

The advantages of this procedure should be apparent from this last example. First, the cancellation of units checks whether the calculation has been set up correctly. If the unwanted units do not cancel, or if the remaining uncanceled units are not the units we are converting to, we know immediately that something is wrong. Second, all the arithmetic is saved until the last step, which means that we pick up the calculator only once and need not write down any intermediate results. The calculator, in fact, is particularly well adapted to chain computations of this sort.

Defined and Measured Conversion Factors

As mentioned earlier, many units have been officially defined in terms of the SI unit for the same quantity. Such definitions are exact and therefore introduce no uncertainty into the calculations. Examples of such defined conversions are

1 furlong = **201.168** m

1 lbm (pound-mass) = **0.453 592 37** kg

Measured Conversion Factors

Other units are defined independently of the SI units, with the conversion factors then established by accurate measurement. Examples are

$$1 \text{ lb (pound-force)} = 4.448\ 222 \text{ N (newton)}$$

$$1 \text{ sidereal year} = 3.155\ 815 \times 10^7 \text{ s}$$

Tables listing these and other conversion factors are found in appendix A. These tables distinguish between defined and measured conversion factors by listing the defined (exact) conversions in boldface type. When doing pencil calculations, this distinction can be made by underlining exact conversions with a wiggly line. Hence

$$1 \text{ scruple} = \mathbf{1.295\ 978\ 2 \times 10^{-3}} \text{ kg}$$

may be handwritten as

$$1 \text{ scruple} = \underset{\sim\sim\sim\sim\sim\sim\sim\sim\sim}{1.295\ 978\ 2 \times 10^{-3}} \text{ kg}$$

but

$$1 \text{ astronomical unit} = 1.496 \times 10^{11} \text{ m}$$

is left as is.

The more common conversion factors have been listed in matrix form. Table A-1, for instance, may be used to find that

$$1 \text{ km} = 0.621\ 371\ 2 \text{ mi}$$
$$1 \text{ ft} = \mathbf{0.304\ 8} \text{ m}$$
$$1 \text{ mi} = \mathbf{63\ 360} \text{ in.}$$

or any of 27 other conversions. The less common units are listed separately, with the conversion factors given only for conversions to the SI unit. If we want to convert from one of these less common units to another uncommon unit, we can still do it without too much difficulty, by the procedure described in example 13-3.

Conversions between Uncommon Units

▶ **Example 13-3: Perches to statute leagues**
Both the perch and the statute league are units of length. Suppose we have a measurement of

$$x = 1\ 854 \text{ perches}$$

which we want to express in statute leagues. The conversion factor between these two units is not given in appendix A; in fact, it has probably never been published anywhere.

We can, however, proceed as follows. Appendix A gives the conversion factors to metres:

$$1 \text{ statute league} = 4.828\,032 \times 10^3 \text{ m}$$
$$1 \text{ perch} = 5.029\,2 \text{ m}$$

Then we can write
$$x = (1\,854 \text{ perches}) \left(\frac{5.029\,2 \text{ m}}{1 \text{ perch}}\right) \left(\frac{1 \text{ statute league}}{4.828\,032 \times 10^3 \text{ m}}\right)$$

Notice again that these conversion fractions have been written so the unwanted units cancel algebraically. This gives
$$x = \frac{(1\,854)(5.092)}{4.828\,032 \times 10^3} \text{ statute leagues}$$
$$x = 1.931 \text{ statute leagues}$$

Four-digit accuracy is retained in this result because the original measurement was given to four-place accuracy. ◄

Dimensional Consistency

The idea of treating units as algebraic quantities can be extended further. If a formula gives the relationship between two quantities, then we can require the same formula to give the relationship between the units for the quantities. This not only reduces the total number of formulas necessary in the sciences, but it also helps prevent algebraic mistakes. This requirement is called *dimensional consistency*.

DEFINITION | *Dimensional consistency is a requirement that a formula relate not only quantities, but also the units for the quantities.*

Suppose, for example, that we need to calculate the volume of a rectangular prism. The formula is
$$V = lwh \qquad (13\text{-}1)$$

where l is the length, w is the width, and h is the height. In the International System, these three quantities would be expressed in metres. If we use U_V to represent the unit of volume, dimensional consistency requires that
$$U_V = \text{m} \cdot \text{m} \cdot \text{m}$$
or
$$U_V = \text{m}^3$$

This formula says that the unit of volume must be metres to the third power, or metres cubed. This unit is commonly called the *cubic metre*, which is all right as long as it is understood to be m³ algebraically. Likewise, we can conclude that the SI unit of area is m², or metres squared.

Dimensional Consistency in Calculations

▶ **Example 13-4: Pouring a concrete floor**
A rectangular floor is to be made of poured concrete. The dimensions of the floor are 26.3 ft by 32.7 ft, with a uniform thickness of 4.0 in. Concrete is ordered by the cubic yard, so the problem is to calculate the required volume in cubic yards.

Dimensionally, a cubic yard is the volume of a perfect cube measuring exactly one yard on each side. This gives

$$U_V = yd \cdot yd \cdot yd$$
$$U_V = yd^3$$

The actual volume is found by using the formula

$$V = lwh$$
$$V = (26.3 \text{ ft})(32.7 \text{ ft})(4.0 \text{ in.})$$

Multiplying just the units together gives

$$ft^2 \cdot in.$$

which is certainly not the unit we seek. We can, however, introduce a few conversion factors directly into the calculation:

$$V = (26.3 \text{ ft})(32.7 \text{ ft})(4.0 \text{ in.})\left(\frac{1 \text{ yd}}{3 \text{ ft}}\right)\left(\frac{1 \text{ yd}}{3 \text{ ft}}\right)\left(\frac{1 \text{ yd}}{36 \text{ in.}}\right)$$

Notice that these conversion factors have been written in such a way as to eliminate the units *ft* and *in.* from the calculation. This leaves

$$V = \frac{(26.3)(32.7)(4.0)}{(3)(3)(36)} yd \cdot yd \cdot yd$$

and, finally,

$$V = 10.6 \text{ yd}^3 \blacktriangleleft$$

The point of this last example is that the validity of the volume formula itself is not based on any particular set of units. It will calculate cubic yards just as well as cubic metres, cubic millimetres, or even cubic perches. Such is the beauty of dimensional consistency.

Sometimes we encounter formulas that are not dimensionally consistent. For example, many empirical equations are not dimensionally consistent: they are valid only with particular combinations of units. There is no special danger with this situation, since such formulas are always accompanied by a detailed list of the units for the quantities represented. They are also often distinguished by the unusual-looking numerical constants in them. Let's look at an example.

Dimensionally Inconsistent Formulas

▶ **Example 13-5: Empirical formula for the inductance of a coil**
The electrical inductance of a coil of wire may be calculated from one of a number of empirical equations, depending on the particular geometry of the coil. Suppose that the coil consists of a single layer of wire (that is, the wire is not wound back over itself), and that the coil is in the shape of a circular cylinder. The inductance is then given by the empirical equation

$$L = 0.039\,69 \; \frac{aN}{l + 0.92a}$$

where L is the inductance in microhenries (μH), a is the coil's diameter in centimetres, l is the coil's length in centimetres, and N is the number of turns of wire.

Now N has no unit at all, since it is just a counting number. This means that the units on the right side of the equation are just

$$\frac{cm^2}{cm + cm}$$

which reduces to units of cm^2/cm, and finally to just plain cm. Yet the formula gives inductance in microhenries, not centimetres. This is dimensional *in*consistency. The use of formulas such as this requires careful attention to the units called for in the accompanying text. ◀

Derived Units

Fortunately, by far the majority of formulas that we encounter *are* dimensionally consistent; in fact, the International System was devised with this consistency in mind. Speed, for instance, is calculated from

$$v = \frac{x}{t}$$

so the SI unit for speed is just the SI unit for distance divided by the SI unit for time:

$$U_v = \frac{m}{s}$$

Using the standard rules for exponents, this may also be written as

$$U_v = \text{m} \cdot \text{s}^{-1}$$

Similarly, the quantity *acceleration* can be defined by

$$a = \frac{\Delta v}{t}$$

where Δv represents a change in speed. The SI unit for acceleration is therefore

$$U_a = \frac{\text{m} \cdot \text{s}^{-1}}{\text{s}}$$

$$= \text{m} \cdot \text{s}^{-1} \cdot \text{s}^{-1}$$

giving
$$U_a = \text{m} \cdot \text{s}^{-2}$$

This procedure applies to any formula that defines a physical quantity. The resulting units are called *derived units*.

DEFINITION | *Derived units* are units that are an algebraic consequence of the dimensional consistency of the formula or equation defining a physical quantity.

Appendix table B-7 lists SI derived units expressed in terms of the seven SI base units. The SI unit for the quantity *moment of inertia*, for instance, is seen to be $\text{kg} \cdot \text{m}^2$.

Derived Units with Special Names

For the sake of convenience, many SI derived units have been given special names. The quantity *force*, for instance, is important in scientific and engineering work, yet it is not listed among the seven SI primary standards and base units. It must therefore be a derived quantity, with a derived unit. In fact, force is defined through Newton's second law of motion:

$$F = ma \qquad (13\text{-}2)$$

where F is the force, m is mass, and a is acceleration. Therefore the unit for force must be

$$U_F = U_m \cdot U_a$$
$$U_F = \text{kg} \cdot \text{m} \cdot \text{s}^{-2}$$

It might be pointed out that the symbol for the quantity should not be confused with the symbol for the unit. In equation (13-2), the symbol m represents mass; the SI unit for this quantity is the kilogram (kg). Similarly, the symbol a represents the quantity acceleration, which, as we saw earlier, has units of $m \cdot s^{-2}$. Here m is the symbol for the SI base unit of length, the metre.

Now, it is just a little cumbersome to express force in units of $kg \cdot m \cdot s^{-2}$; it is even more cumbersome to say it aloud. The International System therefore assigns this combination of base units a special name, the *newton* (N). Thus we can write

$$N = kg \cdot m \cdot s^{-2} \tag{13-3}$$

A number of other SI derived units have special names. The unit for pressure, for instance, is called the *pascal* (Pa). It is given by

$$Pa = N \cdot m^{-2} \tag{13-4}$$

which may be read "newtons per square metre." In terms of the base units, the pascal becomes

$$Pa = kg \cdot m \cdot s^{-2} \cdot m^{-2}$$

which, by combining the m's, simplifies to

$$Pa = kg \cdot m^{-1} \cdot s^{-2} \tag{13-5}$$

The complete list of specially named SI units is given in appendix table B-3.

Using Derived Units in Calculations

▶ **Example 13-6: Fluid power**
A fluid under pressure can be used to produce mechanical motion, for instance, to move a piston in a cylinder. The power developed in such a system is given by the formula

$$P = \mathscr{P}vA$$

where P is the power, $\mathscr{P}$ is the fluid pressure, v is the fluid velocity, and A is the area of the face of the piston.

Suppose that the following measurements have been made:

$$\mathscr{P} = 5.32 \text{ standard atmospheres (atm)}$$

$$v = 4.63 \frac{cm}{s}$$

$$A = 21.6 \text{ cm}^2$$

The problem is to find the power.
Direct substitution gives the following:

$$P = (5.32 \text{ atm})\left(4.63 \frac{cm}{s}\right)(21.6 \text{ cm}^2)$$

The units here combine to give

$$U_P = \text{atm} \cdot \text{cm}^3 \cdot \text{s}^{-1}$$

Referring to appendix tables A-24 and A-25, we see that no such unit of power is listed. Still, dimensional consistency guarantees us that atm · cm³ · s⁻¹ *is* a unit of power, and that therefore it can be converted to any other unit of power we might choose. Let's express the result in the SI power unit, which is the *watt*. We may do so very simply by converting each individual unit to the corresponding SI unit. The *standard atmosphere,* for instance, is found from table A-20 to be

$$1 \text{ atm} = \mathbf{1.013\,25 \times 10^5} \text{ Pa}$$

Similarly, table A-5 tells us that

$$1 \text{ cm}^3 = \mathbf{10^{-6}} \text{ m}^3$$

Then

$$P = (5.32\,\text{atm})(4.63\,\frac{\text{cm}}{\text{s}})(2.16\,\text{cm}^2)\left(\frac{10^{-6}\,\text{m}^3}{1\,\text{cm}^3}\right)\left(\frac{1.013\,25 \times 10^5\,\text{Pa}}{1\,\text{atm}}\right)$$

which reduces to

$$P = 53.9 \text{ Pa} \cdot \text{m}^3 \cdot \text{s}^{-1}$$

Now, if we have done everything right, this should be our answer. But is the combination of units on the right really equivalent to watts? Let's see.

We have already seen that the pascal may be written in terms of base units as

$$\text{Pa} = \text{kg} \cdot \text{m}^{-1} \cdot \text{s}^{-2}$$

(The same information is given in table B-3.) Making this substitution, we have

$$\text{Pa} \cdot \text{m}^3 \cdot \text{s}^{-1} = \text{kg} \cdot \text{m}^{-1} \cdot \text{s}^{-2} \cdot \text{m}^3 \cdot \text{s}^{-1}$$
$$= \text{kg} \cdot \text{m}^2 \cdot \text{s}^{-3}$$

According to table B-3, the watt is given by

$$\text{W} = \text{kg} \cdot \text{m}^2 \cdot \text{s}^{-3}$$

Therefore

$$\text{Pa} \cdot \text{m}^3 \cdot \text{s}^{-1} = \text{W}$$

and our answer may indeed be written

$$P = 53.9 \text{ W} \blacktriangleleft$$

The units with special names make things easy when we talk about so many *pascals* of pressure, or so many *joules* of heat. In calculations, however, these units often have to be expressed in terms of their base units, as we did in the last example.

Finally, we should summarize a few points about dimensional consistency.

> **Rules for Dimensional Consistency**
>
> 1. An SI base unit can never be raised to a fractional power. Expressions like $\sqrt{x}$ where x is in metres, or $\sqrt{t}$ where t is in seconds, are forbidden in dimensionally consistent formulas.
> 2. If two quantities are to be added or subtracted, their units must be the same. Expressions such as $x + t$ do not make sense if x is in metres and t is in seconds.
> 3. A dimensionally consistent formula may not contain transcendental functions of base units. Examples of transcendental functions are the sine, cosine, tangent, logarithm, exponential function, and so on. Therefore expressions such as $\log x$ are not allowable if x is in metres.

Now it may appear that we have ruled out quite a bit, when in fact we have not. The formula for the frequency of oscillation of a simple pendulum, for instance, is

$$f = \frac{1}{2\pi}\sqrt{\frac{g}{l}} \qquad (13\text{-}6)$$

where g is the acceleration of gravity and l is the length of the pendulum. The SI units for these quantities are

$$U_g = \text{m} \cdot \text{s}^{-2}$$
$$U_l = \text{m}$$

The quantity 2π is a pure number with no unit. Then the unit for frequency should be

$$U_f = \sqrt{\frac{U_g}{U_l}}$$
$$= \sqrt{\frac{\text{m} \cdot \text{s}^{-2}}{\text{m}}}$$
$$= \sqrt{\text{s}^{-2}}$$
$$U_f = \text{s}^{-1}$$

This in fact is the correct unit for frequency, and the formula is dimensionally consistent. The point is that we are perfectly justified in taking a square root of m^2, s^{-2}, $\text{m}^2 \cdot \text{s}^{-2}$, A^{-4}, and so on; what we cannot do is take a square root of a base unit itself.

Similarly, the conversion of circular motion to reciprocating motion may be described by

$$y = r \sin(2\pi ft) \qquad (13\text{-}7)$$

where r is the radius of the circle, f is the frequency of rotation, t is time, and y is the linear position of the reciprocating part at time t. Dimensionally, we have

and
$$U_y = U_r \sin(U_f \cdot U_t)$$
$$U_r = \text{m}$$
$$U_f = \text{s}^{-1}$$
$$U_t = \text{s}$$

Then
$$U_y = \text{m} \cdot \sin(\text{s}^{-1} \cdot \text{s})$$
$$= \text{m} \cdot \sin(1)$$
$$U_y = \text{m}$$

which again is correct. We are perfectly free to take transcendental functions of pure numbers; what we cannot do is take such functions of units without sacrificing dimensional consistency.

With these points in mind, we can easily recognize cases of dimensional inconsistency. Remember that only empirical equations are allowed to be dimensionally inconsistent, and then only if the required unit of each quantity is carefully identified. Empirical equations aside, dimensional inconsistency always points to mathematical mistakes. Example 13-7 shows such a case.

Identifying Incorrect Formulas

▶ **Example 13-7: Power from the wind**
Someone claims that the power output of a wind generator may be found from the formula

$$P = \eta \rho r v^3$$

where η (lower-case Greek eta) is the efficiency (a dimensionless quantity), ρ is the density of the air, r is the radius of the vanes, and v is the wind speed. To see whether this formula can be correct, let's check it for dimensional consistency.

$$U_\eta = 1 \qquad \text{(i.e., } \eta \text{ has no unit)}$$
$$U_\rho = \text{kg} \cdot \text{m}^{-3} \qquad \text{(table B-7)}$$
$$U_r = \text{m} \qquad \text{(base unit for length)}$$
$$U_v = \text{m} \cdot \text{s}^{-1} \qquad \text{(table B-7)}$$

Then
$$U_P = \text{kg} \cdot \text{m}^{-3} \cdot \text{m} \cdot (\text{m} \cdot \text{s}^{-1})^3$$
$$= \text{kg} \cdot \text{m}^{-3} \cdot \text{m} \cdot \text{m}^3 \cdot \text{s}^{-3}$$
$$U_P = \text{kg} \cdot \text{m} \cdot \text{s}^{-3}$$

But from table B-3, we see that the SI unit for power is the *watt*, which is expressed in base units as
$$W = \text{kg} \cdot \text{m}^2 \cdot \text{s}^{-3}$$

Since these two sets of units are not identical (one has m while the other has m^2), the original formula is dimensionally inconsistent. Because it is not an empirical equation, it cannot possibly be correct, so it would be fruitless to spend time doing calculations with it. ◄

Now, a formula's dimensional consistency does not *prove* that it is correct; it only implies a high probability that it is right. But a dimensionally *inconsistent* formula is like a flashing red light: such a formula should not be used.

Many other things could be said about units. Units for energy and power are constantly being confused in newspaper and magazine articles. In fact, the two quantities are different and so are their units. Units for mass and weight are totally confused by almost everyone, including many scientists and engineers who should know better. Part of this confusion stems from the fact that the *pound* is a unit for both mass and weight. Yet a pound of mass (lbm) is not the same as a pound of force (lb or sometimes lbf). Similarly, a kilogram is a base unit for mass, but the same name is used for a force unit (kgf). Material that is bought and sold commercially by the "pound" is referred to the lbm rather than to the lbf. Torque that is measured in kilogram-metres is referred to the kgf rather than to the base unit kilogram.

We could go on, but we would soon get into a great deal of physical and chemical theory, and it has not been the purpose of this book to do any such thing. Rather, we have attempted to lay the groundwork of what are too often the "unspoken principles" of a broad cross section of the natural sciences and technologies. It is hoped that you will synthesize your own encounters with technical material with these principles on the limits of observation and the description of things measurable.

Summary

The International System (SI) is the only official system of weights and measures in most of today's world, including the United States. Every other measurement unit is related to an SI unit by an official conversion factor. The seven SI primary standards therefore form the basis of every measurement, regardless of the units.

Once a measurement is made, the result may be rewritten in terms of any valid unit for the measured quantity. The procedure is called *unit conversion,* and is based on treating the unit as an algebraic quantity.

Except for the seven fundamental physical quantities defined by the seven SI primary standards, every physical quantity is defined through an algebraic equation. Such an equation relates the quantity to some combination of the seven fundamental quantities. The same equation relates the unit of the quantity to some combination of the SI base units. When an equation gives a relationship between the quantities as well as between their units, it is *dimensionally consistent.* Except for certain empirical equations, it may be assumed that any valid equation relating physical quantities will be dimensionally consistent.

Units for quantities other than the seven base quantities are *derived units;* each is defined through the same dimensionally consistent equation that defines the quantity it represents. In the International System, many of these derived units are given special names.

In any calculation based on a dimensionally consistent formula, we should take the time to calculate the unit as well. If the calculation does not yield the proper unit for the result, we may assume that a mistake was made somewhere.

REVIEW QUESTIONS

1. What is the importance of the seven SI primary standards?

2. Name the SI base units.

3. Summarize the rules for naming multiples of base units in the International System.

4. What is the only SI base unit that already contains a metric prefix?

5. How many fundamentally different physical quantities are there? Name them.

6. Outline the procedure for converting a measurement result from one unit to another.

7. How does a defined conversion factor differ from a measured conversion factor?

8. What is meant by dimensional consistency?

9. What is the advantage of using dimensionally consistent formulas?

10. What type of formula is frequently not dimensionally consistent?

11. What precaution against misuse should you take when writing a dimensionally inconsistent formula?

12. What is a derived unit?

13. What does dimensional consistency have to do with derived units?
14. Give examples of some of the SI derived units with special names, and their equivalents in terms of SI base units.
15. What are the rules for writing dimensionally consistent formulas?
16. How may the dimensional consistency requirement be used to identify incorrect formulas?

EXERCISES

1. Perform the following unit conversions. Refer to table B-4 for the definitions of the SI prefixes.

   a. 1 231 m to kilometres
   b. 0.003 27 m to millimetres
   c. 3 987 kg to Megagrams
   d. 9.4×10^9 s to Gigaseconds
   e. 6.83×10^7 Ω to Megohms
   f. 4.11×10^{-5} A to microamperes
   g. 0.036 MT to teslas
   h. 2.11×10^{-8} s to nanoseconds
   i. 5.4×10^{-10} F to picofarads
   j. 4.8×10^4 μm to metres

   [*Answers:* (a) 1.231 km; (c) 3.987 Mg; (f) 41.1 μA; (g) 36 000 T]

2. Perform the following unit conversions. The conversion factors may be found in appendix A.

   a. 763.14 ft to metres
   b. 4 954 km to miles
   c. 3.009 1 m^2 to square inches
   d. 0.065 417 hyl to kilograms
   e. 185.4 lbm to kilograms
   f. 55.00 USA liquid gallons to liters
   g. 1 262.137 Pa to millimetres of mercury
   h. 983.21 cusec-hours to liters
   i. 0.050 3 in^2 to circular mils
   j. 10.21 parsecs to light-years

   [*Answers:* (a) 232.61 m; (c) 4 664.1 in^2; (e) 84.10 kg; (h) $1.002\ 29 \times 10^8$ ℓ]

3. A rectangular swimming pool measures 23.2 ft by 73.9 ft. The bottom slopes continuously from a depth of 8.22 ft at one end to 4.18 ft at the other. Calculate the volume of water held by the pool in (a) USA liquid gallons, and (b) liters. (*Answer:* 301 000 ℓ)

4. The gauge pressure at a given depth of water may be found from

$$\mathscr{P} = (\text{depth}) \cdot (\text{density}) \cdot g$$

where g is the acceleration due to gravity. The density of water at 19.6°C is 0.998 285 g/cm^3. Find the pressure at a depth of 98.28 ft of fresh water at 30° north latitude: (a) in pascals, and (b) in pounds per square inch.

5. The power output of a motor is related to its torque output by

$$P = 2\pi f T$$

where f is the frequency of rotation. The formula is dimensionally consistent. Find the SI unit for frequency.

6. The force due to air resistance on a moving object is

$$F = cv^2$$

where v is the object's speed and c is a quantity sometimes called the drag coefficient.
   a. Express the unit of c in terms of the SI base units.
   b. At a speed of 53.2 mi/h, a certain car experiences a force of 478 lbf of air resistance. Find the value of c in SI units.

7. The electrical energy "consumed" by an appliance in a time t can be found from

$$E = Pt$$

where P is the appliance's power rating. A certain pump motor requires 998 W of electrical power. Find the energy it uses in 3.27 h (a) in joules, (b) in kilowatt-hours, (c) in Btu's, (d) in foot-pounds. (*Answer:* 11 100 Btu)

8. An object's mass may be found by multiplying its density by its volume:

$$\text{mass} = (\text{density}) \cdot (\text{volume})$$

Steel has a density of 7.86 g/cm^3.
   a. What is the mass, in lbm, of a coil of steel sheet 0.032 in. thick, 3.21 ft wide, and 1 272 ft long?
   b. What is the coil's mass in kilograms? (*Answer:* 2 430 kg)

APPENDIX A

Tables of Conversion Factors

This appendix lists the more common English and metric conversions in matrix form. To use each matrix, locate the unit in the row heading at the left, then read its equivalent in terms of the appropriate column heading. In the first table, for instance, we may read

1 foot = **0.304 8** metres.

Figures listed in **boldface** are exact conversions. The others are generally given to seven-place accuracy.

Very large, very small, and relatively uncommon units are also listed, but separately from the matrixes. In most cases, these units are given only in terms of their SI equivalents. These units may, however, be converted to any other non-SI units by combining two conversion factors. To convert fathoms to international nautical miles, for instance, we may write

$$1 \text{ fathom} = (1.828\ 8 \text{ m}) \left(\frac{1 \text{ natuical mile}}{1.852 \times 10^3 \text{ m}} \right)$$

which reduces to

1 fathom = **9.874 73** $\times$ **10^{-4}** nautical miles.

See chapter 13 for a more detailed discussion of the use of unit conversion factors.

Table A-1. *Length and Distance: English and Metric Conversions*

	cm	m	km	in	ft	mi
1 centimetre =	1	10^{-2}	10^{-5}	0.393 700 8	3.280 840 $\times 10^{-2}$	6.213 712 $\times 10^{-6}$
1 metre =	100	1	10^{-3}	39.370 08	3.280 840	6.213 712 $\times 10^{-4}$
1 kilometre =	10^5	1 000	1	3.937 008 $\times 10^4$	3 280.840	0.621 371 2
1 inch =	2.54	0.025 4	2.540 $\times 10^{-5}$	1	8.333 333 $\times 10^{-2}$	1.578 283 $\times 10^{-5}$
1 foot =	30.48	0.304 8	3.048 $\times 10^{-4}$	12	1	1.893 939 $\times 10^{-4}$
1 statute mile =	160 934.4	1 609.344	1.609 344	63 360	5 280	1

Table A-2. *Length and Distance: Less Common Units*

Length Unit	Size in Metres	Comments
angstrom (Å)	10^{-10}	
astronomical unit (AU)	1.496×10^{11}	equal to the mean distance from the earth to the sun
Bohr radius	$5.291\,77 \times 10^{-11}$	used in atomic physics
chain, Gunter's or surveyor's	20.116 84	100 links or 4 rods
chain, Ramden's or engineer's	30.48	100 ft
cubit, Palestinian	0.641	ancient
cubit, Roman	0.444	ancient
cubit, modern	0.457 2	
fathom	1.828 8	6 ft
femtometer (fm)	10^{-15}	also called a *fermi*
fermi	10^{-15}	used in nuclear physics; same as a *femtometer*
foot, Cape	0.314 858 1	
foot, geodetic Cape	0.314 855 575 16	
foot, South African geodetic	0.304 797 265 4	also called *English foot* by land surveyors
foot, US survey	0.304 800 609 6	
furlong	201.168	1/8 mi or 220 yd
hand	0.101 6	
iron	5.3×10^{-4}	used in shoe manufacture
league, international nautical	5.556×10^{3}	
league, UK nautical	$5.559\,552 \times 10^{3}$	
league, statute	$4.828\,032 \times 10^{3}$	
light-year	$9.463\,7 \times 10^{15}$	the distance light travels in 1 year
ligne	6.35×10^{-4}	used for making buttons; equal to 1/40 in.
link, Gunter's or surveyor's	0.201 168 4	
link, Ramden's or engineer's	0.304 8	same as a foot
micrometre (μm)	10^{-6}	same as a micron
micron	10^{-6}	same as a micrometre
mil	2.54×10^{-5}	0.001 in.; sometimes called a *thou*
millimetre (mm)	10^{-3}	
nanometre (nm)	10^{-9}	
nautical mile, international	1.852×10^{3}	
nautical mile, UK	$1.853\,184 \times 10^{3}$	
nautical mile, USA	1.852×10^{3}	
nautical mile, telegraph	$1.855\,32 \times 10^{3}$	
pace	0.762	
parsec	$3.085\,7 \times 10^{16}$	used in astronomy
perch	5.029 2	there is also an area unit by this name; same as a *pole*

Continued

Table A-2. *Continued*

Length Unit	Size in Metres	Comments
pica	$4.217\ 518 \times 10^{-3}$	used in printing; approximately 12 points
picometre (pm)	10^{-12}	
point	$3.514\ 598 \times 10^{-4}$	used in printing
pole	**5.029 2**	there is also an area unit by this name; same as a *perch*
rod	5.029 210	25 links or **16.5** ft
rood, Cape	3.778 297 2	**12** times the Cape foot
rood, geodetic Cape	3.778 266 9	
Siegbahn unit	10^{-13}	
skein	**109.728**	**360** ft
span	0.228 6	
yard	**0.914 4**	**3** ft or **36** in

Note: Boldface type indicates exact conversions.

Table A-3. *Area: English and Metric Conversions*

	cm²	m²	in²	ft²
1 square centimetre =	1	10^{-4}	0.155 000 31	$1.076\ 391 \times 10^{-3}$
1 square metre =	10^4	1	1 550.003 1	10.763 91
1 square inch =	**6.451 6**	$\mathbf{6.451\ 6} \times \mathbf{10^{-4}}$	1	$6.944\ 444 \times 10^{-3}$
1 square foot =	929.030 4	$9.290\ 304 \times 10^{-2}$	**144**	1

Table A-4. *Area: Less Common Units*

Area Unit	Size in Square Metres	Comments
acre	4.04686×10^3	43560 ft^2
are	100	
barn	10^{-28}	100 square femtometers
circular mil (cmil)	5.067075×10^{-10}	the area of a circle 1 mil in diameter; equal to $\pi/4$ square mils, or $7.853982 \times 10^{-5} \text{ in}^2$
hectare	10^4	
morgen	8.56532×10^3	
perch	25.2929	there is also a length unit by this name; same as a *pole*
pole	25.2929	there is also a length unit by this name; same as a *perch*
rood, UK	1.011715×10^3	approximately ¼ acre
square mil	6.4516×10^{-10}	one-millionth of a square inch
square mile (mi²)	2.589988×10^6	640 acres; also called a *section*
square yard (yd²)	0.836127 36	
township	9.323957×10^7	36 mi^2

Note: Any unit in the length table may be squared for form a square unit. The appropriate conversion factor in such cases may be found by squaring the converion factor found in the length table. For example, 1 square skein = $(109.728 \text{ m})^2 = 1.20402 \times 10^4 \text{ m}^2$.

Table A–5. *Volume (or Capacity): English and Metric Conversions*

	cm³	ℓ	m³	in³	gallon	ft³
1 cubic centimetre =	1	10^{-3}	10^{-6}	$6.102\,374 \times 10^{-2}$	$2.641\,721 \times 10^{-4}$	$3.531\,467 \times 10^{-5}$
1 liter* =	10^3	1	10^{-3}	61.023 74	0.264 172 1	$3.531\,467 \times 10^{-2}$
1 cubic metre =	10^6	10^3	1	61 023.74	264.172 1	35.314 67
1 cubic inch =	16.387 064	$1.638\,706\,4 \times 10^{-2}$	$1.638\,706\,4 \times 10^{-5}$	1	$4.329\,004 \times 10^{-3}$	$5.787\,037 \times 10^{-4}$
1 gallon, USA liquid =	3 785.412	3.785 412	$3.785\,412 \times 10^{-3}$	231	1	0.133 680 6
1 cubic foot =	28 316.85	28.316 85	$2.831\,685 \times 10^{-2}$	1 728	7.480 519	1

*Redefined in 1964 (Resolution 6, 12th CGPM) to equal exactly **1 000** cm³. Previously, the liter was defined as the volume of **1** kg of water at its maximum density, equal to 1 000.027 cm³.

Table A-6. *Volume (or Capacity): Less Common Units*

Volume Unit	Size in Cubic Metres	Comments
acre-foot	$1.233\ 482 \times 10^3$	an area of 1 acre covered to a depth of 1 ft
barrel, dry	$0.115\ 627\ 12$	7 056 in^3; used for dry commodities other than cranberries
barrel, dry, for cranberries	$9.547\ 103 \times 10^{-2}$	5 826 in^3
barrel, liquid	$1.589\ 873 \times 10^{-1}$	42 USA gallons
board foot (fbm)	$2.359\ 737 \times 10^{-3}$	144 in^3, or $1/12$ ft^3
bushel, UK	$3.636\ 872 \times 10^{-2}$	
bushel, USA	$3.523\ 907 \times 10^{-2}$	4 USA pecks, or 32 USA dry quarts
cord	$3.624\ 557$	128 ft^3
cubic decimetre (dm^3)	10^{-3}	1 liter
cubic yard (yd^3)	$0.764\ 554\ 9$	27 ft^3
cup, UK	$2.841\ 306 \times 10^{-4}$	
cup, USA	$2.365\ 882 \times 10^{-4}$	
cusec-hour	$1.019\ 407 \times 10^2$	
drachm, UK fluid	$3.551\ 633 \times 10^{-6}$	
dram, USA fluid	$3.696\ 691 \times 10^{-6}$	$1/8$ USA fluid ounce
fifth	$7.570\ 824 \times 10^{-4}$	used for alcoholic spirits; equal to $1/5$ USA liquid gallon, or 46.2 in^3
fluid ounce, UK	$2.841\ 306 \times 10^{-5}$	
fluid ounce, USA	$2.957\ 353 \times 10^{-5}$	
gallon, Canadian liquid	$4.546\ 122 \times 10^{-3}$	same as the British imperial gallon; equal to the volume of 10 lb water at 62°F
gallon, USA liquid	$3.785\ 412 \times 10^{-3}$	4 USA liquid quarts
gallon, UK	$4.546\ 09 \times 10^3$	
gallon, USA dry	$4.404\ 884 \times 10^{-3}$	4 USA dry quarts
gill, UK	$1.420\ 653 \times 10^{-4}$	
gill, USA	$1.182\ 941 \times 10^{-4}$	0.25 USA fluid ounces
lambda	10^{-9}	
leaguer	$0.577\ 353\ 4$	
milliliter (mℓ)	10^{-6}	1 cm^3
minim	$5.919\ 39 \times 10^{-8}$	480 USA fluid ounces
morgen-foot	$2.610\ 71 \times 10^3$	an area of 1 morgen covered to a depth of 1 ft
ounce, UK fluid	$2.841\ 306 \times 10^{-5}$	
ounce, USA fluid	$2.957\ 353 \times 10^{-5}$	
peck, UK	$9.092\ 18 \times 10^{-3}$	
peck, USA	$8.809\ 768 \times 10^{-3}$	8 USA dry quarts
pint, UK	$5.682\ 613 \times 10^{-4}$	
pint, USA dry	$5.506\ 105 \times 10^{-4}$	

Continued

Table A-6. *Continued*

Volume Unit	Size in Cubic Metres	Comments
pint, USA liquid	$4.731\ 765 \times 10^{-4}$	
quart, USA dry	$1.101\ 221 \times 10^{-3}$	**2** USA dry pints
quart, USA liquid	$9.463\ 529 \times 10^{-4}$	**2** USA liquid pints, **4** USA cups, or **32** USA fluid ounces
register ton	$2.831\ 685$	used for capacity of ships
stere	1	used for firewood
tablespoon, UK	$1.420\ 653 \times 10^{-5}$	
tablespoon, USA	$1.478\ 676 \times 10^{-5}$	
teaspoon, UK	$4.735\ 51 \times 10^{-6}$	
teaspoon, USA	$4.928\ 922 \times 10^{-6}$	

Note: Any unit in the length table may be cubed to form a cubic unit. For example, 1 cubic mile = $(1\ 609.344\text{ m})^3$, which equals $4.168\ 182 \times 10^9\text{ m}^3$.

Table A-7. Plane Angle: English and Metric Conversions

	radian	°	'	"	r	gon
1 radian =	1	180/π or 57.295 78	3 437.747	206 264.8	0.159 154 9	63.661 98
1 degree =	π/180 or 1.745 329 × 10^{-2}	1	60	3 600	2.777 778 × 10^{-3}	1.111 111 1
1 minute =	2.908 882 × 10^{-4}	1.666 667 × 10^{-2}	1	60	4.629 630 × 10^{-5}	1.851 852 × 10^{-2}
1 second =	4.848 137 × 10^{-6}	2.777 778 × 10^{-4}	1.666 667 × 10^{-2}	1	7.716 049 × 10^{-7}	3.086 420 × 10^{-4}
1 revolution =	2π or 6.283 185	360	2.16 × 10^4	1.296 × 10^6	1	400
1 gon* =	π/200 or 1.570 796 × 10^{-2}	0.9	54	3 240	2.5 × 10^{-3}	1

*Also called a *grade*; based on a standard of 1 right angle = 100 gon.

Table A–8. *Time: English and Metric Conversions*

	s	min	h	dy	y
1 second =	1	$1.666\ 667 \times 10^{-2}$	$2.777\ 778 \times 10^{-4}$	$1.157\ 407 \times 10^{-5}$	$3.168\ 876 \times 10^{-8}$
1 minute =	60	1	$1.666\ 667 \times 10^{-2}$	$6.944\ 444 \times 10^{-4}$	$1.901\ 326 \times 10^{-6}$
1 hour =	3 600	60	1	$4.166\ 667 \times 10^{-2}$	$1.140\ 795 \times 10^{-4}$
1 day, mean solar =	86 400.002	518.400 0	24.000 00	1	$2.737\ 909 \times 10^{-3}$
1 year,* tropical =	$3.155\ 693 \times 10^7$	525 948.8	8 765.814	365.242 2	1

*Currently shrinking at a rate of 5.3 ms/y

Table A-9. *Time: Less Common Units*

Time Unit	Equivalent in Seconds	Comments
day, sidereal	$8.616\ 409 \times 10^4$	23 h, 56 min, and 4.09 s; based on the apparent motion of the stars rather than the sun
month, anomalistic	$2.380\ 71 \times 10^6$	27.554 6 mean solar days
month, nodical	$2.351\ 136 \times 10^6$	27.212 22 mean solar days
month, sidereal	$2.360\ 591 \times 10^6$	27.321 66 mean solar days
month, synodic	$2.551\ 443 \times 10^6$	29.530 59 mean solar days; sometimes called the *phase month of the moon*
month, tropical	2.552 427	29.530 40 mean solar days
year, Muslim	$3.061\ 728 \times 10^7$	average; a cycle of **30** Muslim years contains **19** years with **354** days and **11** leap years with **355** days, for an average of 354.366 7 mean solar days per Muslim year
year, sidereal	$3.155\ 815 \times 10^7$	365.255 9 mean solar days, or 366.256 2 sidereal days

Table A-10. *Speed: English and Metric Conversions*

	$\dfrac{cm}{s}$	$\dfrac{m}{s}$	$\dfrac{km}{h}$	$\dfrac{ft}{s}$	$\dfrac{ft}{min}$	$\dfrac{mi}{h}$
1 centimetre per second =	1	10^{-2}	3.6×10^{-2}	$3.280\,840 \times 10^{-2}$	1.968 504	$2.236\,936 \times 10^{-2}$
1 metre per second =	100	1	3.6	3.280 840	196.850 4	2.236 936
1 kilometre per hour =	27.777 78	0.277 777 8	1	0.911 344 4	54.680 66	0.621 371 2
1 foot per second =	30.48	0.304 8	1.097 28	1	60	0.681 818 2
1 foot per minute =	0.508	5.08×10^{-3}	$1.828\,8 \times 10^{-2}$	$1.666\,667 \times 10^{-2}$	1	$1.136\,364 \times 10^{-2}$
1 mile* per hour =	44.704	0.447 04	1.609 344	1.466 667	88	1

*Statute

Table A-11. *Speed: Less Common Units*

Speed Unit	Equivalent in Metres per Second	Comments
foot per hour	$8.466\ 667 \times 10^{-5}$	
inch per second	$\mathbf{2.54 \times 10^{-2}}$	
kilometre per minute	1.666 667	
kilometre per second	$\mathbf{10^3}$	
knot, international	0.514 444 4	1 international nautical mile per hour
knot, UK	0.514 773 3	
knot, USA	0.514 444	
mile per minute	**26.822 4**	60 mi/h
mile per second	$\mathbf{1.609\ 344 \times 10^3}$	

Table A–12. *Acceleration: English and Metric Conversions*

	$\dfrac{cm}{s^2}$	$\dfrac{m}{s^2}$	$\dfrac{in.}{s^2}$	$\dfrac{ft}{s^2}$	g
1 centimetre per second, per second* =	1	10^{-2}	0.393 700 8	$3.280\,840 \times 10^{-2}$	$1.019\,716 \times 10^{-3}$
1 metre per second, per second =	10^2	1	39.370 08	3.280 84	0.101 971 6
1 inch per second, per second =	2.54	2.54×10^{-2}	1	$8.333\,333 \times 10^{-2}$	$0.372\,971\,4$
1 foot per second, per second =	30.48	0.304 8	12	1	$3.108\,095 \times 10^{-2}$
1 standard free-fall acceleration =	980.665	9.806 65	386.088 6	32.174 05	1

*Also called a *gal*, after Galileo.

Table A-13. *Acceleration: Less Common Units*

Unit of Acceleration	Squared Equivalent in Metres per Second	Comments
dyne per gram	**10^{-2}**	1 cm/s^2
gal	**10^{-2}**	1 cm/s^2
kilometre per hour, per second	0.277 777 8	
mile per hour, per second	**0.447 04**	
newton per kilogram (N/kg)	1	1 m/s^2
pound per slug	**0.304 8**	1 ft/s^2

Table A-14. *Mass: English and Metric Conversions*

	g	kg	tonne	lbm	slug
1 gram =	1	10^{-3}	10^{-6}	$2.204\ 623 \times 10^{-3}$	$6.852\ 177 \times 10^{-5}$
1 kilogram =	10^{3}	1	10^{-3}	2.204 623	$6.852\ 177 \times 10^{-2}$
1 tonne =	10^{6}	10^{3}	1	2 204.623	48.521 77
1 pound mass, avoirdupois* =	453.592 37	0.453 592 37	$4.535\ 923\ 7 \times 10^{-4}$	1	$3.108\ 095 \times 10^{-2}$
1 slug =	14 593.902 9	14.593 902 9	$1.459\ 390\ 29 \times 10^{-2}$	32.174 05	1

*This should not be confused with the force (or weight) unit called the avoirdupois pound. An object with a *weight* of 1 lb will weigh a different amount if taken to a different latitude or a different altitude. This *weight* change can easily amount to 2 or 3 percent. Since variations of this magnitude are clearly unacceptable in commerce, goods are always bought or sold by *mass* rather than *weight*. An object with a *mass* of 1 lbm will have this same mass regardless of its location.

At places where the gravitational acceleration has a value of 9.806 65 m/s² (the so-called standard free-fall acceleration), a 1-lb mass will weigh exactly 1 lb. Since such places are rare, it is important to make a careful distinction between the units of mass and the units of force (or weight). For example, calibration of spring scales for commercial use is done so the scale indicates *mass* rather than *weight*. If such a scale is moved to another location where the gravitational acceleration is different, it must be recalibrated.

Table A-15. *Mass: Less Common Units*

Mass Unit	Equivalent in Kilograms	Comments
atomic mass unit (amu)	$1.660\ 531 \times 10^{-27}$	
carat, metric	2×10^{-4}	used for precious stones
drachm	$3.887\ 934\ 6 \times 10^{-3}$	same as a troy dram
dram, avoirdupois	$1.771\ 845\ 195\ 312\ 5 \times 10^{-3}$	$1/16$ of an avoirdupois ounce-mass
dram, troy or apothecary	$3.887\ 934\ 6 \times 10^{-3}$	60 grains; same as a drachm
grain	$6.479\ 891 \times 10^{-5}$	
hundredweight, long	$50.802\ 345\ 44$	112 lbm
hundredweight, short	$45.359\ 237$	100 lbm
hyl	$9.806\ 65$	sometimes called a *metric slug*
megagram (Mg)	10^3	tonne, or metric ton
microgram (μg)	10^{-9}	10^{-6} g
milligram (mg)	10^{-6}	10^{-3} g
ounce mass, avoirdupois	$2.834\ 952\ 312\ 5 \times 10^{-2}$	$1/16$ lbm
ounce mass, troy or apothecary	$3.110\ 347\ 68 \times 10^{-2}$	480 grains
pennyweight	$1.555\ 174 \times 10^{-3}$	
pound mass (lbm), avoirdupois	$0.453\ 592\ 37$	
pound mass, troy or apothecary	$0.373\ 241\ 721\ 6$	12 troy ounce-masses
quarter	$12.700\ 59$	2 stones
quintal	100	
scruple, apothecary	$1.295\ 978\ 2 \times 10^{-3}$	
stone	$6.350\ 293$	
ton, long	$1.016\ 046\ 908\ 8 \times 10^3$	2 240 lbm
ton, metric	10^3	1 tonne, or Mg
ton, short	$9.071\ 847\ 4 \times 10^2$	2 000 lbm, or 20 short hundredweights

Table A-16. *Force, Weight: English and Metric Conversions*

	N	kgf	lb	ton
1 newton =	1	0.101 971 6	0.224 808 9	$1.124\ 045 \times 10^{-4}$
1 kilogram-force =	9.806 65	1	2.204 623	$1.102\ 311 \times 10^{-3}$
1 pound =	4.448 222	0.453 592 37	1	5×10^{-4}
1 ton-force =	8 896.444	907.184 7	2 000	1

Table A-17. *Force, Weight: Less Common Units*

Force Unit	Equivalent in Newtons	Comments
dyne	10^{-5}	
gram-force (gf)	$9.806\ 65 \times 10^{-3}$	
kilopond	9.806 65	1 kgf
kip	$4.448\ 221\ 615\ 260\ 5 \times 10^3$	1 000 lb
ounce-force, avoirdupois	0.278 013 85	
poundal	0.138 254 954 376	
pound, avoirdupois	4.448 221 615 260 5	
ton-force, metric	$9.806\ 65 \times 10^3$	

Table A-18. *Temperature Conversion Formulas: English and Metric Conversions*

to find ↓	from °C	from K	from °F	from °R
degrees Celsius	—	$K - 273.15$	$\frac{5}{9}(F - 32)$	$\frac{5}{9}R - 273.15$
kelvin*	$C + 273.15$	—	$\frac{5}{9}(F + 459.67)$	$\frac{5}{9}R$
degrees Fahrenheit	$\frac{9}{5}C + 32$	$\frac{9}{5}K - 459.67$	—	$R - 459.67$
degrees Rankine	$\frac{9}{5}C + 459.67$	$\frac{9}{5}K$	$F + 459.67$	—

Notes: Formulas in the table give exact conversions.

Also see table 5-4, page 69, for a list of Fahrenheit–Celsius equivalents. For temperatures between about 200°F and 800°F, or 100°C and 400°C, the Fahrenheit temperature is approximately *twice* the Celsius temperature.

*Sometimes erroneously called "degrees Kelvin."

Table A-19. Density (Mass per Unit Volume): English and Metric Conversions

	$\dfrac{g}{cm^3}$	$\dfrac{kg}{m^3}$	$\dfrac{lbm}{in^3}$	$\dfrac{lbm}{ft^3}$	$\dfrac{slug}{ft^3}$
1 gram per cm³ =	1	10^3	$3.612\,729 \times 10^{-2}$	$62.427\,96$	$1.940\,320$
1 kilogram per m³ =	10^{-3}	1	$3.612\,729 \times 10^{-5}$	$6.242\,796 \times 10^{-2}$	$1.940\,320 \times 10^{-3}$
1 lbm per in³ =	$27.679\,905$	$27\,679.905$	1	$1\,728.000$	$53.707\,88$
1 lbm per ft³ =	$1.601\,846 \times 10^{-2}$	$16.018\,463$	$5.787\,037 \times 10^{-4}$	1	$3.108\,095 \times 10^{-2}$
1 slug per ft³ =	$0.515\,378\,8$	$515.378\,8$	$1.861\,924 \times 10^{-2}$	$32.174\,05$	1

Tables of Conversion Factors

Table A–20. *Pressure: English and Metric Conversions*

	Pa	$\dfrac{kgf}{m^2}$	mm Hg	$\dfrac{lb}{ft^2}$	$\dfrac{lb}{in^2}$	in. Hg	atm
1 pascal =	1	0.101 971 6	7.500 616 $\times 10^{-3}$	2.088 544 $\times 10^{-2}$	1.450 377 $\times 10^{-4}$	2.952 998 $\times 10^{-4}$	9.869 233 $\times 10^{-7}$
1 kilogram-force per m² =	9.806 65	1	7.355 592 $\times 10^{-2}$	0.204 816 1	1.422 334 $\times 10^{-3}$	2.895 902 $\times 10^{-3}$	9.678 415 $\times 10^{-5}$
1 millimetre of mercury at 0°C =	133.322 39	13.595 10	1	2.784 496	1.933 678 $\times 10^{-2}$	3.937 007 $\times 10^{-2}$	1.315 790 $\times 10^{-3}$
1 pound per ft² =	47.880 258	4.882 428	0.359 131 4	1	6.944 444 $\times 10^{-3}$	1.413 903 $\times 10^{-2}$	4.725 413 $\times 10^{-4}$
1 pound per in² =	6 894.757 2	703.069 6	51.714 93	144	1	2.036 020	6.804 596 $\times 10^{-2}$
1 inch of mercury at 32°F =	3 386.389	345.315 6	25.4	70.726 20	0.491 154 2	1	3.342 106 $\times 10^{-2}$
1 standard atmosphere =	101 325	10 332.275	760	2 116.217	14.695 95	29.921 26	1

Table A-21. *Pressure: Less Common Units*

Unit of Pressure	Equivalent in Pascals	Comments
bar	10^5	10^6 dynes/cm^2
centimetre of mercury at 0°C	$1.333\ 223\ 9 \times 10^3$	
centimetre of water at 4°C	98.063 8	
dyne per cm^2	0.1	
foot of water at 39.2°F	$2.988\ 98 \times 10^3$	
inch of mercury at 60°F	$3.376\ 85 \times 10^3$	
inch of water at 39.2°F	$2.490\ 82 \times 10^2$	
inch of water at 60°F	$2.488\ 4 \times 10^2$	
kilogram-force per cm^2	$9.806\ 65 \times 10^4$	
kilopond per cm^2	$9.806\ 65 \times 10^4$	
kip per in^2	$6.894\ 757 \times 10^6$	
metre of water at 4°C	$9.806\ 38 \times 10^3$	
millibar	100	10^3 dynes/cm^2
newton per m^2	1	
pièze	10^3	
poundal per ft^2	1.488 164	
poundal per in^2	214.296	
ton-force per in^2	$1.378\ 95 \times 10^7$	
torr	$1.333\ 223\ 7 \times 10^2$	

Table A-22. *Energy, Work, Heat: English and Metric Conversions*

	J	cal	kw·h	ft·lb	Btu	hp·h
1 joule =	1	0.239 005 7	2.777 777 × 10^{-7}	0.737 562 1	9.484 514 × 10^{-4}	0.372 506 × 10^{-6}
1 calorie (thermochemical) =	4.184	1	1.162 222 × 10^{-6}	3.085 960	3.968 321 × 10^{-3}	1.558 57 × 10^{-6}
1 kilowatt-hour =	3.6 × 10^{6}	860 420.7	1	2.655 224 × 10^{6}	3.414.425	1.341 02
1 foot-pound =	1.355 818	0.324 048 3	3.766 161 × 10^{-7}	1	1.285 927 × 10^{-3}	5.050 50 × 10^{-7}
1 British thermal unit (thermochemical) =	1 054.350	251.995 8	2.928 751 × 10^{-4}	777.648 8	1	3.927 52 × 10^{-4}
1 horsepower-hour =	2.684 52 × 10^{6}	6.416 16 × 10^{5}	0.745 700	1.980 00 × 10^{6}	2 546.14	1

Table A-23. *Energy, Work, Heat: Less Common Units*

Energy Unit	Equivalent in Joules	Comments
Btu, International Steam Table	$1.055\ 04 \times 10^3$	
Btu, mean	$1.055\ 87 \times 10^3$	
Btu, thermochemical	$1.054\ 350\ 264\ 488 \times 10^3$	
Btu (39°F)	$1.059\ 67 \times 10^3$	
Btu (60°F)	$1.054\ 68 \times 10^3$	
calorie (cal), International Steam Table	$4.186\ 8$	
calorie, mean	$4.190\ 02$	
calorie (15°C)	$4.185\ 80$	
calorie (20°C)	$4.181\ 90$	
dyne-centimetre	10^{-7}	same as an erg
electron-volt (ev)	$1.602\ 191\ 7 \times 10^{-19}$	
erg	10^{-7}	
foot-poundal	$4.214\ 011\ 0 \times 10^{-2}$	
joule, international of 1948	$1.000\ 165$	
kgf-metre	$9.806\ 65$	
kilocalorie (kcal)		10^3 of the corresponding calorie unit
kilopond-metre	$9.806\ 65$	same as a kgf-metre
kilowatt-hour, international of 1948	$3.600\ 59 \times 10^6$	
liter-atmosphere	$1.013\ 28 \times 10^2$	
million electron-volt (Mev)	$1.602\ 191\ 7 \times 10^{-13}$	
therm	$1.055\ 06 \times 10^8$	
thermie	$4.185\ 5 \times 10^6$	
ton of TNT, nuclear equivalent	4.20×10^9	
watt-hour	3.60×10^3	
watt-second	1	

Table A-24. *Power (Energy per Unit Time): English and Metric Conversions*

	W	kW	$\dfrac{\text{cal}}{\text{s}}$	$\dfrac{\text{Btu}}{\text{h}}$	$\dfrac{\text{ft} \cdot \text{lb}}{\text{s}}$	hp
1 watt =	1	10^{-3}	0.239 005 7	3.414 425	0.737 562 2	1.341 022 $\times 10^{-3}$
1 kilowatt =	10^3	1	239.005 7	3 414.425	737.562 2	1.341 022
1 calorie (thermochemical) per second =	4.184	4.184×10^{-3}	1	14.285 95	3.085 960	5.610 836 $\times 10^{-3}$
1 Btu (thermochemical) per hour =	0.292 875 1	2.928 751 $\times 10^{-4}$	6.999 883 $\times 10^{-2}$	1	0.216 013 6	3.927 520 $\times 10^{-4}$
1 foot-pound per second =	1.355 817 9	1.355 817 9 $\times 10^{-3}$	0.324 048 3	4.629 338	1	1.818 182 $\times 10^{-3}$
1 horsepower =	745.699 87	0.745 699 87	178.226 5	2 546.136	550	1

Table A-25. *Power (Energy per Unit Time): Less Common Units*

Unit of Power	Equivalent in Watts	Comments
Btu (International Steam Table) per hour	0.293 071 1	
Btu (thermochemical) per minute	17.572 504	
Btu (thermochemical) per second	$1.054\ 350\ 264\ 488 \times 10^3$	
calorie (International Steam Table) per hour	1.163×10^{-3}	
calorie (thermochemical) per minute	$6.973\ 333 \times 10^{-2}$	
calorie (thermochemical) per second	4.184	
cheval vapeur	735.499	also called the *metric horsepower*
erg per second	10^{-7}	
foot-pound per hour	$3.766\ 161 \times 10^{-4}$	
foot-pound per minute	$2.259\ 697 \times 10^{-2}$	
frigerie	1.162 639	
horsepower, boiler	$9.809\ 50 \times 10^3$	
horsepower, electric	746	
horsepower, metric	735.499	also called a *cheval vapeur*
horsepower, UK	745.7	
horsepower, water	746.043	
joule per second	1	
ton (12 000 Btu per hour)	$3.516\ 853 \times 10^3$	used in refrigeration
ton (13 440 Btu per hour)	$3.938\ 876 \times 10^3$	used in refrigeration
watt, international of 1948	1.000 165	

Table A-26. Torque: English and Metric Conversions

	N·m	kgf·m	kgf·cm	lb·in	lb·ft
1 newton-metre =	1	0.101 971 6	10.197 16	8.850 748	0.737 562 1
1 kilogram-force-metre =	9.806 65	1	100	86.796 19	7.233 014
1 kilogram-force-centimetre =	9.80665×10^{-2}	10^{-2}	1	0.867 961 9	7.233014×10^{-2}
1 pound-inch =	0.112 984 8	1.152124×10^{-2}	1.152 124	1	8.333333×10^{-2}
1 pound-foot	1.355 818	0.138 255 0	13.825 50	12	1

Table A-27. *Absolute Viscosity: Less Common Units*

Unit of Absolute Viscosity	Equivalent in Newton-second per Square Metre, or Pascal-second
centipoise	10^{-3}
kilogram per metre-second	1
lbm per foot-second	1.488 163 9
poise	0.1
slug per foot-second	4.788 025 8

Table A-28. *Kinematic Viscosity: Less Common Units*

Unit of Kinematic Viscosity	Equivalent in Square Metres per Second
centistoke	10^{-6}
square foot, per second	$9.290\ 304 \times 10^{-2}$
stoke	10^{-4}

Note: There are other measures of viscosity (such as the SUS and the SAE numbers) that are not true units and therefore cannot be related to the SI viscosity unit through a simple conversion factor.

Table A-29. *Electrical Resistivity: Less Common Units*

Resistivity Unit	Equivalent in Ohm-metres
ohm-centimetre	10^{-2}
ohm-cmil, per foot	$1.662\ 426\ 1 \times 10^{-9}$
ohm-foot	0.304 8
ohm-inch	2.54×10^{-2}

Table A-30. *Thermal Conductivity: English and Metric Conversions*

	$\dfrac{W}{m \cdot K}$	$\dfrac{cal}{cm \cdot s \cdot C°}$	$\dfrac{kcal}{m \cdot hr \cdot C°}$	$\dfrac{Btu \cdot in.}{ft^2 \cdot hr \cdot F°}$	$\dfrac{Btu}{ft \cdot hr \cdot F°}$
1 watt per metre-Kelvin =	1	2.390 057 × 10^{-3}	0.860 420 7	6.938 113	0.578 176 1
1 calorie per centimetre-second-C° =	418.4	1	360	2 902.907	241.908 9
1 kilocalorie per metre-hour-C° =	1.162 222	2.777 778 × 10^{-3}	1	8.063 630	0.671 969 3
$\dfrac{Btu \cdot in.}{ft^2 \cdot hr \cdot F°}$	0.144 131 4	3.444 823 × 10^{-4}	0.124 013 6	1	8.333 333 × 10^{-2}
$\dfrac{Btu}{ft \cdot hr \cdot F°}$	1.729 577	4.133 788 × 10^{-3}	1.488 163	12	1

Table A-31. *Heat of Transformation, Latent Heat: Less Common Units*

Unit of Latent Heat	Equivalent in Joules per Kilogram
Btu (thermochemical), per lbm	$2.324\ 444 \times 10^3$
calorie (thermochemical), per gram	4.184×10^3
foot-pound, per slug	$9.290\ 304 \times 10^{-5}$
joule per gram	10^3

Table A-32. *Magnetic Flux: Less Common Units*

Magnetic Flux Unit	Equivalent in Webers (Wb)
emu	10^{-8}
kiloline	10^{-5}
line	10^{-8}
maxwell	10^{-8}

Table A-33. *Magnetic Flux Density: Less Common Units*

Magnetic Flux Density Unit	Equivalent in Teslas (T)
gamma	10^{-9}
gauss	10^{-4}
kiloline per square inch	$1.550\ 003 \times 10^{-2}$
line per square centimetre	10^{-4}
milligauss	10^{-7}
weber per square metre	1

Table A-34. *Magnetic Field Strength: Less Common Units*

Magnetic Field Strength Unit	Equivalent in Ampere-turns per Metre
ampere-turn per centimetre	100
ampere-turn per inch	39.370 08
esu	$2.654\ 418 \times 10^{-9}$
gilbert	79.577 47
oersted	79.577 47

APPENDIX B

The International System of Units (SI)

The original metric system, a product of the French Revolution of 1789, consisted of a reformed and standardized system of weights and measures. Throughout history, this "metric system" has varied quite a bit from place to place and has been subject to various political forces. In 1960, it finally became completely standardized by international treaty, and thus evolved into the Système Internationale d'Unités (SI), commonly called the international system in English-speaking countries.

Table B-1. *The SI Base Units*

	SI Base Unit		
Quantity	Name	International Symbol	Definition (CGPM)
length	metre	m	The metre is the length equal to 1 650 763.73 wavelengths in vacuum of the radiation corresponding to the transition between the levels $2p_{10}$ and $5d_5$ of the krypton-86 atom [11th CGPM (1960), Resolution 6].
mass	kilogram	kg	The kilogram is the mass of the international prototype kilogram recognized by the CGPM and in the custody of the Bureau International des Poids et Mesures, Sèvres, France [1st CGPM (1889)].
time	second	s	The second is the duration of 9 192 631 770 periods of the radiation corresponding to the transition between the two hyperfine levels of the ground state of the cesium-133 atom [13th CGPM (1967), Resolution 1].
electric current	ampere	A	The ampere is the constant current that, if maintained in two straight parallel conductors of infinite length, of negligible circular cross section, and placed 1 metre apart in vacuum would produce between these conductors a force equal to 2×10^{-7} N per metre of length [CGPM (1946), Resolution 2, approved by the 9th CGPM (1948)].
thermodynamic temperature	kelvin	K	The kelvin, unit of thermodynamic temperature, is the fraction 1/273.16 of the thermodynamic temperature of the triple point of water [13th CGPM (1967), Resolution 4].

Continued

Table B-1. *Continued*

Quantity	SI Base Unit		
	Name	International Symbol	Definition (CGPM)
luminous intensity	candela	cd	The candela is the luminous intensity, in the perpendicular direction of a surface of 1/600 000 m^2 of a blackbody at the temperature of freezing platinum under a pressure of 101 325 N/m^2 [13th CGPM (1967), Resolution 5].
amount of substance	mole	mol	The mole is the amount of substance of a system that contains as many elementary entities as there are atoms in 0.012 kg of carbon-12 [14th CGPM (1971), Resolution 3].

Table B-2. *The SI Supplementary Units*

Quantity	SI Unit		
	Name	Symbol	Definition
plane angle	radian	rad	The radian is the plane angle between two radii of a circle that cut off on the circumference an arc equal in length to the radius.
solid angle	steradian	sr	The steradian is the solid angle that, having its vertex in the centre of a sphere, cuts off an area of the surface of the sphere equal to that of a square with sides of length equal to the radius of the sphere.

Table B-3. *The SI Derived Units Having Special Names*

			SI Unit	
Quantity	Name	Symbol	Expression in Terms of Other Units	Expression in Terms of SI Base Units
frequency	hertz	Hz		s^{-1}
force	newton	N		$m \cdot kg \cdot s^{-2}$
pressure	pascal	Pa	N/m^2	$m^{-1} \cdot kg \cdot s^{-2}$
energy, work, quantity of heat	joule	J	$N \cdot m$	$m^2 \cdot kg \cdot s^{-2}$
power, radiant flux	watt	W	J/s	$m^2 \cdot kg \cdot s^{-3}$
quantity of electricity, electric charge	coulomb	C	$A \cdot s$	$s \cdot A$
electric potential, potential difference, electromotive force	volt	V	W/A	$m^2 \cdot kg \cdot s^{-3} \cdot A^{-1}$
capacitance	farad	F	C/V	$m^{-2} \cdot kg^{-1} \cdot s^4 \cdot A^2$
electric resistance	ohm	Ω	V/A	$m^2 \cdot kg \cdot s^{-3} \cdot A^{-2}$
conductance	siemens	S	A/V	$m^{-2} \cdot kg^{-1} \cdot s^3 \cdot A^2$
magnetic flux	weber	Wb	$V \cdot s$	$m^2 \cdot kg \cdot s^{-2} \cdot A^{-1}$
magnetic flux density	tesla	T	Wb/m^2	$kg \cdot s^{-2} \cdot A^{-1}$
inductance	henry	H	Wb/A	$m^2 \cdot kg \cdot s^{-2} \cdot A^{-2}$
luminous flux	lumen	lm		$cd \cdot sr$*
illuminance	lux	lx		$m^{-2} \cdot cd \cdot sr$*

*In these expressions, the steradian (sr) is treated as a base unit for solid angle. For a source radiating in all directions from a point, the solid angle is 4π steradians.

Table B-4. *Preferred SI Prefixes*

Factor		Factor in Words	SI Prefix	SI Symbol
1 000 000 000 000	or 10^{12}	trillion*	tera-	T
1 000 000 000	or 10^{9}	billion*	giga-	G
1 000 000	or 10^{6}	million	mega-	M
1 000	or 10^{3}	thousand	kilo-	k
0.001	or 10^{-3}	thousandth	milli-	m
0.000 001	or 10^{-6}	millionth	micro-	μ
0.000 000 001	or 10^{-9}	billionth*	nano-	n
0.000 000 000 001	or 10^{-12}	trillionth*	pico-	p
0.000 000 000 000 001	or 10^{-15}	—	femto-	f
0.000 000 000 000 000 001	or 10^{-18}	—	atto-	a

Note: These prefixes may be applied to any SI unit, whether base, supplementary, or derived. Occasionally, they are also used with certain non-SI units. For example, 1 024 000 m = 1.024 × 10^6 m = 1.024 Mm.

*U.S. usage. In most of the world, 10^9 is called a "milliard," and 10^{12} (a million million) is called a "billion." The European billion is therefore 1 000 times as large as the U.S. billion. To prevent ambiguity, the words "billion" and "trillion" should be avoided when at all possible.

Table B-5. *Other SI Prefixes*

Factor		Factor in Words	SI Prefix	SI Symbol
100	or 10^{2}	hundred	hecto-	h
10	or 10^{1}	ten	deca-	da
0.1	or 10^{-1}	tenth	deci-	d
0.01	or 10^{-2}	hundredth	centi-	c

Note: The use of these prefixes is relatively uncommon and thus should be avoided when possible.

Table B-6. *Non-SI Units That May Be Used with the International System*

Quantity	Name and Symbol	Value in Terms of the SI
mass	metric ton (t)	1 t = 1 Mg = 1 000 kg
plane angle	degree (°)	1 ° = $\pi/180$ rad
plane angle	minute (')	1 ' = $\pi/10\,800$ rad
plane angle	second (")	1 " = $\pi/648\,000$ rad
plane angle	gon (gon)	1 gon = $\pi/200$ rad
time	minute (min)	1 min = 60 s
time	hour (h)	1 h = 3.6 ks = 3 600 s
time	day (d)	1 d = 86.4 ks = 86 400 s
time	week (wk)	1 wk = 604.8 ks = 604 800 s
volume fluid	liter (ℓ)	1 ℓ = 1 dm^3 = 0.001 m^3

Note: These units find widespread use because of their convenient size, and the CGPM does not expect them to be eliminated in the foreseeable future.

Table B-7. *SI Derived Units That May Be Expressed in Terms of SI Base and Supplementary Units*

Quantity	SI Unit Name	Symbol
acceleration	metre per second squared	m/s^2
angular acceleration	radian per second squared	rad/s^2
angular momentum	kilogram-metre squared per second	$kg \cdot m^2/s$
angular velocity	radian per second	rad/s
area	square metre	m^2
coefficient of linear expansion	1 per kelvin	K^{-1}
concentration (of amount of substance)	mole per cubic metre	mol/m^3
density	kilogram per cubic metre	kg/m^3
diffusion coefficient	metre squared per second	m^2/s
electric current density	ampere per square metre	A/m^2
exposure rate (ionizing radiation)	ampere per kilogram	A/kg
kinematic viscosity	metre squared per second	m^2/s
luminance	candela per square metre	cd/m^2
magnetic field strength	ampere per metre	A/m
magnetic moment	ampere-metre squared	$A \cdot m^2$
mass flow rate	kilogram per second	kg/s
mass per unit area	kilogram per square metre	kg/m^2
mass per unit length	kilogram per metre	kg/m
molality	mole per kilogram	mol/kg
molar mass	kilogram per mole	kg/mol
molar volume	cubic metre per mole	m^3/mol
moment of inertia	kilogram-metre squared	$kg \cdot m^2$
moment of momentum	kilogram-metre squared per second	$kg \cdot m^2/s$
momentum	kilogram-metre per second	$kg \cdot m/s$
radioactivity (disintegration rate)	1 per second	s^{-1}
rotational frequency	1 per second	s^{-1}
specific volume	cubic metre per kilogram	m^3/kg
speed	metre per second	m/s
velocity	metre per second	m/s
volume	cubic metre	m^3
wave number	1 per metre	m^{-1}

Table B-8. SI Derived Units That May Be Expressed in Terms of SI Base and Supplementary Units as Well as SI Derived Units with Special Names

Quantity	SI Unit		Expression in Terms of SI Base Units and SI Supplementary Units
	Name	Symbol	
absorbed dose	joule per kilogram	J/kg	$m^2 \cdot s^{-2}$
coefficient of heat transfer	watt per metre squared-kelvin	W/m$^2 \cdot$ K	$kg \cdot s^{-3} \cdot K^{-1}$
conductivity	siemens per metre	S/m	$m^{-3} \cdot kg^{-1} \cdot s^3 \cdot A^2$
dielectric polarization	coulomb per square metre	C/m^2	$m^{-2} \cdot s \cdot A$
displacement	coulomb per square metre	C/m^2	$m^{-2} \cdot s \cdot A$
dynamic viscosity	pascal second	Pa $\cdot$ s	$m^{-1} \cdot kg \cdot s^{-1}$
electric charge density	coulomb per cubic metre	C/m^3	$m^{-3} \cdot s \cdot A$
electric dipole moment	coulomb-metre	C $\cdot$ m	$m \cdot s \cdot A$
electric field strength	volt per metre	V/m	$m \cdot kg \cdot s^{-3} \cdot A^{-1}$
energy density	joule per cubic metre	J/m^3	$m^{-1} \cdot kg \cdot s^{-2}$
entropy	joule per kelvin	J/K	$m^2 \cdot kg \cdot s^{-2} \cdot K^{-1}$
exposure (ionizing radiation)	coulomb per kilogram	C/kg	$kg^{-1} \cdot s \cdot A$

Quantity	Unit	Symbol	Expression
heat capacity	joule per kelvin	J/K	$m^2 \cdot kg \cdot s^{-2} \cdot K^{-1}$
heat flux density	watt per square metre	W/m²	$kg \cdot s^{-3}$
magnetic dipole moment	weber-metre	Wb·m	$m^3 \cdot kg \cdot s^{-2} \cdot A^{-1}$
molar energy	joule per mole	J/mol	$m^2 \cdot kg \cdot s^{-2} \cdot mol^{-1}$
molar entropy	joule per mole-kelvin	J/mol·K	$m^2 \cdot kg \cdot s^{-2} \cdot K^{-1} \cdot mol^{-1}$
molar heat capacity	joule per mole-kelvin	J/mol·K	$m^2 \cdot kg \cdot s^{-2} \cdot K^{-1} \cdot mol^{-1}$
moment of force	newton-metre	N·m	$m^2 \cdot kg \cdot s^{-2}$
permeability	henry per metre	H/m	$m \cdot kg \cdot s^{-2} \cdot A^{-2}$
permittivity	farad per metre	F/m	$m^{-3} \cdot kg^{-1} \cdot s^4 \cdot A^2$
radiant intensity	watt per steradian	W/sr	$m^2 \cdot kg \cdot s^{-3} \cdot sr^{-1}$
reluctance	1 per henry	H⁻¹	$m^{-2} \cdot kg^{-1} \cdot s^2 \cdot A^2$
resistivity	ohm-metre	Ω·m	$m^3 \cdot kg \cdot s^{-3} \cdot A^{-2}$
specific energy	joule per kilogram	J/kg	$m^2 \cdot s^{-2}$
specific entropy	joule per kilogram-kelvin	J/kg·K	$m^2 \cdot s^{-2} \cdot K^{-1}$
specific heat capacity	joule per kilogram-kelvin	J/kg·K	$m^2 \cdot s^{-2} \cdot K^{-1}$
specific latent heat	joule per kilogram	J/kg	$m^2 \cdot s^{-2}$
surface charge density	coulomb per square metre	C/m²	$m^{-2} \cdot s \cdot A$
surface tension	newton per metre	N/m	$kg \cdot s^{-2}$
thermal conductivity	watt per metre-kelvin	W/m·K	$m \cdot kg \cdot s^{-3} \cdot K^{-1}$
torque	newton-metre	N·m	$m^2 \cdot kg \cdot s^{-2}$

INDEX

Absolute differences. *See* Method of absolute differences
Absolute pressure: 218
Absorption spectrophotometer: 223
Accuracy: converting to uncertainty, 54–55; definition of, 47, 51; implied uncertainty, 55–56; instrument, 46–47, 51, 54; in meter scale, 48–49; in multirange instruments, 49; and precision, 52–54; and systematic error, 50
Ampere, primary standard of: 6
Angular coordinate. *See* Polar graph
Arithmetic mean: 28–30; agreement with median, 28; definition of, 28; determination, 28
Average. *See* Arithmetic mean

Bar graph, construction: 74–75
Base units. *See* Units
Bellows gauges: 219
Beta gauge: 225

CRT. *See* Cathode-ray tube indicator
Calibration: 10–12; definition of, 10; use in thermometers, 4
Calibration curve: 174
Calorimeters: 224
Candela, primary standard of: 6
Capacitance, sensor: calculating, 211–212; definition of, 210; lag time relationship, 214; and length measurement, 215–216; uncertainty principle, 212–215
Cartesian coordinate system: 152; compared with Polar system, 157
Cathetometer: 215

Cathode ray tube indicator: 201, 203
Center of gravity: definition, 128; determination, 128–139. *See also* Method of absolute differences
Chaotic error: 40–41
Chemical measurement. *See* Chemical sensors
Chemical sensors: 222–223. *See also* names of individual instruments
Circle graph, construction: 76–77
Cloud chambers: 224–225; bubble chamber, 225
Coefficient of friction: 135
Compass direction: 157, 160; PPI radar scope, 159, 162
Compound units. *See* Derived units
Concentric circles: 153
Confidence level: 37–38; definition of, 37; true value in, 37–38
Conversion factors: defined, 235; measured, 235–236; tables, 249–277
Conversions: compound units, 234–235; between uncommon units, 236–237; unit, 232–234
Count, of set of numbers: 2

d'Arsonval meter movement: 190–191; applications, 194–195; damping, 191–192; lag time in, 193–194
Data table: applications, 61–62; organization, 62–63; unequal uncertainties in, 64–65; usefulness, 63. *See also* Fundamental tables; Statistical tables
Dead zone: 199–201; definition of, 199; determination, 201
Defined units: 9; definition of, 10

287

Dependent variable: 77
Derived units: in calculations, 241–243; definition of, 10, 239–240; relation to dimensional consistency, 239–240; special names, 240–241
Descartes, René: 152
Deviations: 31–36; calculating, 32; definition of, 31. *See also* Maximum deviation; Mean absolute deviation; Standard deviation
Deviations from the mean: *See* Deviations
Dial-type indicator: 188; hydraulic, 199; pneumatic, 199
Diaphragm capsule: 218
Digital indicators: 201–202
Dimensional consistency: in calculations, 238–239; definition of, 237; inconsistency, 239; in identifying incorrect formulas, 244–245; rules for, 243
Discarding data point: 29–30
Discrimination: 51–52
Display unit. *See* Indicator
Dosimeter: 224

Electrical resistors: 16
Electron diffraction pattern: 227
Electron microscope: 227
Empirical equations: accuracy, 132; advantages, 121; definitions, 120; linear, 126–127; nonlinear, 139–143; uncertainty in, 122; use, 122. *See also* Log-log graph; Method of absolute difference; Slope; Straight line; Y intercept
Equation: 120. *See also* Empirical equation
Error: 15–16; acceptable, 15; propagation of, 96–97. See also Chaotic error; Measurement error; Parallax error; Probable error; Random error; Scale-limited error
Extrapolation: definition of, 69; linear, 70

Flowcharts: 57; measurement procedure, 58
French curve: 82–83
Functional graphs: construction, 78–79; definition of, 77
Functional table: 65, extrapolation with, 69–70; interpolation with, 66–68
Fundamental physical quantity: 232

Gas chromotography: 223
Gauge pressure: 218
Geiger-Müller tubes: 224
Graph: advantages and disadvantages, 73–76; combinations, 86–91; definition of, 73; drawing uncertainties in, 84; grids, 91; and precision, 74. *See also* Bar graph; Cartesian coordinate system; Circle graph; Dependent variable; Functional graphs; Independent variable; Interpolation; Log-log graphs; Logarithmic graphs; Polar graphs; Semilog graphs; Straight line graph; Uncertain graph
Grids: 91

IPTS. *See* International Practical Temperature Scale
Inch: 9
Independent variable: 77; in constructing graph, 88–91
Index of refraction: 143–146
Indicator: 187–211; application, 204–205; definition of, 187; properties, 192–193. *See also* d'Arsonval meter movement; Dial-type indicator; Digital indicators; Micrometer indicators; Vernier
Indirect measurement, determining uncertainty in: 99–101
Instrument accuracy. *See* Accuracy
Instrument discrimination. *See* Discrimination
Interferometer: 216
International Committee on Weights and Measures: 5
International Practical Temperature Scale: 8–9; fixed points of, 9; Kelvin, 6
International System: 5; base units, 7, 232; primary standards, 6, 231; relation to standards, 231; tables, 278–285
Interpolation: alternative to direct calculation, 69; definition of, 66; in empirical equations, 122; in graphs, 79–82, 85; linear, 67–69, 80; nonlinear, 80–83
Inverted bell sensors: 219
Ionization chambers: 224; proportional counter, 224; semiconductor junction sensors, 224

Kelvin. *See* International Practical Temperature Scale
Kilogram, primary standard of: 6–7

Lag time: 192–195; of d'Arsonval movement, 193–194; and sensor capacitance, 214
Laser: 216
Length measurement, instruments for: 215–216. *See also* names of individual instruments
Linear interpolation. *See* Interpolation
Log-log graphs: 175, 177–178; and empirical

equations, 178-183. *See also* Semilog graphs
Logarithm: 168-170; determination, 170-172; use of calculator, 170-171
Logarithmic graphs: 167-184; definition, 168; drawing, 172-174; log paper, 175-176; logarithmic graph paper, 174-175. *See also* Log-log graphs; Semilog graphs
Lost motion. *See* Dead zone

Mathematics, role of: 1
Maximum deviation: 32-33, definition of, 33; use in tolerancing, 32
Mean. *See* Arithmetic mean
Mean absolute deviation: calculating, 34; definition of, 33
Mean deviation. *See* Mean absolute deviation
Measurement: 2; direct, 99-101; indirect, 99-100; limitations, 14; reliability, 17. *See also* Chemical sensors; Conversions; Length measurement; Photographs; Pressure measurement; Radiation sensors; Speed measurement; Temperature measurement
Measurement error: 14-24; affected by instrument accuracy, 50; definition of, 15; systematic error, 17; uncertainty in, 16, 25
Measurement uncertainty. *See* Uncertainty
Median: 27-30; agreement with mean, 28-30; definition of, 27; determination, 27-28
Mercury barometers: 218
Meter. *See* Metre
Meter damping: 191-192
Meter scale: 195; accuracy of, 48-49; linear meter scale, 195; nonlinear, 195-196; Parallax error, 196-199
Method of absolute differences: 128-129; in equation of the best straight line, 130-131; and nonlinear data, 143; use, 131-132; use when Y intercept is zero, 133-134
Method of least squares: 128
Metre: primary standard of, 6; Treaty of the, 7; as unit of length in SI, 231
Metric system: 5. *See also* International System
Micrometer indicators: 188; caliper, 188, 212
Mole, primary standard of: 6
Moving coil. *See* d'Arsonval movement

National Bureau of Standards: 7-8
Nonlinear data: 135-138; and absolute differences, 143; curves, 136-137. *See also* Nonlinear equations
Nonlinear equations, derivations: 139-143
Nonlinear interpolation. *See* Interpolation
Null measurement: advantages, 206; definition of, 205; use, 206
Number, physical significance of: 2

Odometer: 215
Orifice manometer: 173
Oscilloscope: 203

PPI radar scope. *See* Compass direction
Parallax error: 196-199; definition of, 197; flattened pointer, 199; minimization, 198; mirrored scale, 198
Pen-type recorders: drum-type recorder, 204; strip chart recorder, 203, 204; X-Y recorder, 204
Peter's formula: 35-36
Photoelectric sensor: 223; absorption spectrum, 223
Photographs as sensors: 226-229; applications, 226-229; electron diffraction patterns, 227; electron microscope, 227
Plan position indicator radar scope. *See* Compass direction
Pneumatic transmitter: 218
Polar coordinate system: compared with Cartesian system, 158; definition of, 152-153. *See also* Polar graph
Polar graph: angular coordinate, 160, 162; curved spokes, 161; drawing, 154-156; plotting data, 153-154; use, 153. *See also* Compass direction
Precision: and accuracy, 52-56; of common dimensioning instruments, 54; definition, 52; in graphs, 74
Pressure measurement, instruments for: 218-219; absolute pressure, 216; gauge pressure, 216. *See also* names of individual instruments
Pressure transducers: 219
Primary standards. *See* Standards
Probable error: 16-17; defined as uncertainty, 25
Propagation of uncertainty: 103-106; definition of, 104; direct calculation, 109; examples, 105-106; in a function of a single measured variable, 105; in a product of two measured variables, 106-107; in a ratio of two measured variables, 108. *See also* Significant digits
Psychrometer: 161, 164

Quantities, fundamental physical: 6

Radiation measurement. *See* Radiation sensors
Radiation sensors: instruments, 223–226; use, 225–226. *See also* names of individual instruments
Random error: 20–22, 41; and accuracy, 50; specimen variation, 20–22; with systematic error, 50–51; uncertainty, 25–27; uncontrolled variables in, 20–21
Reliability: 17
Risetime: 192

SI. *See* International System
Scale-limited error: 39–40
Scintillation counter: 224; photomultiplier tube, 224
Secondary standard. *See* Standards
Semilog graphs: 175, 177–179. *See also* Log-log graphs
Second, primary standard of: 6
Sensor: definition of, 186, 209; design requirements, 209–210. *See also* Capacitance, sensor; Chemical sensors; Length measurement; Photographs; Pressure measurement; Radiation sensors; Speed measurement
Significant digits: 113–116; in calculation, 114–115; definition of, 114; rule, 115; rule limitations, 115–116
Significant zero: 56
Slope: definition, 123; determination, 123–125
Specimen variation: 20–22; in maximum deviation, 33
Spectrometer: 220, 223; diffraction grating, 125
Spectrophotometer: 223
Speed measurement, instruments for: 216–218. *See also* names of individual instruments
Speedometer: 217
Squint test: 85–86
Standard deviation: 34–37; calculating, 36; definition of, 34; Peter's formula, 35–36; true value in, 37
Standards: definition and use, 3–4; laboratory, 5; primary, 4, 231; relation to International System, 231; secondary, 4
Statistical table: 66
Straight line: equation, 123; graph, 127. *See also* Center of gravity; Log-log graph; Method of absolute difference; Slope

Strain: 216
Strain gauge: 216
Strip chart recorders: 203, 204
Stroboscope: 217
Subscripts: 31
Summation notation: 30–31
Systematic error: 17–20; and accuracy, 50; corrections for, 25; definition of, 17; detection of, 19; minimization of, 18; with random error, 50–51; sources, 18; uncertainty, 25

Tables. *See* Data tables
Tachometer: 217
Temperature measurement: instruments for, 219–221; limits of precision, 221. *See also* names of individual instruments
Thermometer: accuracy, 46–47; gas, 220; ultrasonic, 220; use with standards, 3–4
Time constant: 193, and sensor capacitance, 214
Tolerance, definition of: 33
Tracers, radioactive: 225
Transducers: 216
Traveling microscope: 216
Treaty of the Metre. *See* Metre, Treaty of the
True value: 16, 22; as arithmetic mean, 28, 30; best estimations, 25; in confidence level, 38; as median, 27; in standard deviation, 37
Tube sensors: 218

Ultrasonic depth meters: 216
Uncertain graph, determining uncertainty from: 101–103
Uncertainty: calculations with, 97, 110; in calculated quantity, 107; conversion of instrument accuracy to, 54; definition, 25; determining, 30–37; deviations, 30–37; in empirical equations, 122; graphs, 84–85; implied, 54–55; measurement, 56–57; in random error, 26–27; rounding off, 64; in systematic error, 25–26. *See also* Propagation of uncertainty
Uncertainty principle: 212–214
Uncontrolled variable: 20–21; acting at random, 27; in mean absolute deviation, 41; in standard deviation, 41
Units: 1–13, 231–246; base, 231–246. *See also* Conversions; Defined units; Derived units; Dimensional consistency; Dimensional inconsistency

Vacuum gauge: 218; ionization, 218; radioactive sources used in, 226; thermal, 218

Vernier: 188

Voltmeter: accuracy, 48–50; applications, 51; examples of use, 47, 48, 51, 52

Wavelength standards: 143–146

X-Y recorder: 204, 205

Y intercept: definition, 125–126; as zero, 132–133